Medical Nihilism

Medical Nihilism

Jacob Stegenga

OXFORD
UNIVERSITY PRESS

OXFORD
UNIVERSITY PRESS

Great Clarendon Street, Oxford, OX2 6DP,
United Kingdom

Oxford University Press is a department of the University of Oxford.
It furthers the University's objective of excellence in research, scholarship,
and education by publishing worldwide. Oxford is a registered trade mark of
Oxford University Press in the UK and in certain other countries

© Jacob Stegenga 2018

The moral rights of the author have been asserted

First Edition published in 2018
Impression: 8

Published in the United States of America by Oxford University Press
198 Madison Avenue, New York, NY 10016, United States of America

British Library Cataloguing in Publication Data
Data available

Library of Congress Control Number: 2017955745

ISBN 978-0-19-874704-8

Printed and bound by
CPI Group (UK) Ltd, Croydon, CR0 4YY

For those who suffer

Contents

Part II. Methods

Part III. Evidence and Values

Acknowledgments

Nancy Cartwright has had the most pronounced influence on this book—the rigor of her own work has been a constant spur, and her pastoral care has fostered this book in countless inarticulable ways. Denis Walsh helped to provide the time and the confidence to begin writing. I am grateful to Alex Broadbent, Jonathan Fuller, Maël Lemoine, and Anna Vaughn for their close reading, questions, and criticisms of draft versions of large parts of this manuscript at various stages of development.

Many people have offered valuable commentary and discussion on particular chapters. For this I am grateful to Anna Alexandrova, Hanne Andersen, Richard Ashcroft, Peggy Battin, Ken Bond, Frédéric Bouchard, Craig Callender, Martin Carrier, Nancy Cartwright, Hasok Chang, Rachel Cooper, Heather Douglas, Marc Ereshefsky, Martyn Evans, Luis Flores, Leslie Francis, Fermin Fulda, James Gardner, Beatrice Golomb, Sara Green, Marta Halina, Jon Hodge, Bennett Holman, Jeremy Howick, Philippe Huneman, Phyllis Illari, Stephen John, Saana Jukola, Aaron Kenna, Brent Kious, James Krueger, Adam La Caze, Eric Martin, Leah McClimans, Boaz Miller, Elijah Millgram, David Moher, Peter Momtchiloff, Robert Northcott, Rune Nyrup, Barbara Osimani, Wendy Parker, Anya Plutynski, Dasha Pruss, Gregory Radick, Isaac Record, Federica Russo, Simon Schaffer, Samuel Schindler, Jonah Schupbach, Miriam Solomon, Jan Sprenger, Georg Starke, Veronica Strang, James Tabery, Eran Tal, Mariam Thalos, Aleksandra Traykova, Jonathan Tsou, Denis Walsh, Sarah Wieten, Torsten Wilholt, John Worrall, and Alison Wylie. I am also grateful to audiences at numerous universities and conferences. A risk of having so many interlocutors over so many years is that I am liable to forget to thank all of them here. I am truly sorry if I have done so.

Dr. Brent Kious (psychiatry) participated in one of my philosophy of medicine seminars, and Dr. Benjamin Lewis (psychiatry) allowed me to shadow his clinical work with patients at the University of Utah Neuropsychiatric Institute. Dr. Beatrice Golomb (internal medicine) introduced me to some of the problems of assessing harms of medical interventions. Dr. Dick Zoutman (infectious diseases) first emphasized to me the practical difficulty of amalgamating diverse evidence. Dr. Luis Flores (psychiatry), health policy analyst Ken Bond, and epidemiologist David Moher provided written commentary on particular chapters. I am grateful to Dr. Samuel Brown (intensive care), Dr. Howard Mann (radiology), Dr. Jeffrey Botkin (pediatrics and research ethics), and Dr. Willard Dere (personalized medicine) for valuable discussion on several general themes from the book.

Parts of this book have appeared in previous publications. A version of Chapter 2 and some of Chapter 3 was published as "Effectiveness of Medical Interventions" in *Studies in History and Philosophy of Biological and Biomedical Sciences* (hereafter

Studies C) (2015a). A version of Chapter 5 was published as "Down with the Hierarchies" in *Topoi* (2014). A version of Chapter 6 was published as "Is Meta-Analysis the Platinum Standard?" in *Studies C* (2011). A version of Chapter 7 was published as "Herding QATs: Quality Assessment Tools for Evidence in Medicine" in *Classification, Disease, and Evidence: New Essays in Philosophy of Medicine* (2015b). A version of Chapter 8 was published as "Measuring Effectiveness" in *Studies C* (2015c). A version of Chapter 9 was published as "Hollow Hunt for Harms" in *Perspectives on Science* (2016). Appendix 4 was developed with Jan Sprenger in "Three Arguments for Absolute Outcome Measures" in an article in *Philosophy of Science* (forthcoming). Parts of Chapters 8 and 12 are published as "Drug Regulation and the Inductive Risk Calculus" in *Exploring Inductive Risk* (2017). I have extensively reworked, excised, and added material from these articles to improve and clarify their arguments and to unify them as parts of a coherent whole. Most of Chapters 1, 4, 10, 11, 12, and the technical material in the appendices are new.

I began this book while I was a Fellow of the Banting Postdoctoral Fellowships Program, administered by the Government of Canada, and held at the Institute for the History and Philosophy of Science and Technology at the University of Toronto. I am grateful for this generous support. It is a nice historical accident that my research for those two years was supported by a program named after Frederick Banting, the scientist who discovered the biological basis of, and effective treatment for, type 1 diabetes. Banting's great achievement, which saved so many lives and mitigated profound suffering, has raised my standard for what we can hope for from medicine, and I use this standard in the arguments that follow.

I wrote most of this book while at University of Utah, University of Victoria, and then University of Cambridge—these institutions provided generous time for me to devote to writing. I made the final touches during a fellowship at the Institute of Advanced Study in Durham University. This idyllic setting of riverside paths, medieval stone roads, and wooded trails was conducive to testing a claim made by Hippocrates, an early medical nihilist and the symbolic parent of western medicine: walking is our best medicine.

Parlance

A book about medical research written by a philosopher of science necessarily relies on technical jargon from both medicine and philosophy. I try to explain such terminology where appropriate.

When referring to pharmaceuticals I usually use their scientific names and in parentheses I note their trade names. The latter are usually more familiar. Many drugs have multiple trade names but I usually list only one.

Occasionally I employ simple notation from probability theory, in which $P(X)$ means "the probability of X" and $P(X|Y)$ means "the probability of X given Y." For example, $P(\text{rain today}) = 0.6$ means "the probability of rain today is 0.6," or "there is a 60% chance that it will rain today," and $P(\text{rain today}|\text{I'm in Seattle}) = 0.8$ means "given that I'm in Seattle, the probability of rain today is 80%."

I use the following abbreviations.

ADHD   attention deficit hyperactivity disorder

BMJ   *British Medical Journal* (former name, now just referred to as *BMJ*)

CDER   Center for Drug Evaluation and Research

CML   chronic myelogenous leukemia

DSM   Diagnostic and Statistical Manual of Mental Disorders

EBM   evidence-based medicine

EMA   European Medicines Agency

FDA   Food and Drug Administration

HAMD   Hamilton Rating Scale for Depression

NHLBI   National Heart, Lung, and Blood Institute

NICE   National Institute for Clinical Excellence

NIH   National Institutes of Health

NIMH   National Institute of Mental Health

NNT   number needed to treat

PPAR   peroxisome proliferator activated receptor

QAT   quality assessment tool

RCT   randomized controlled trial

RD   risk difference
RR   relative risk
RRR   relative risk reduction
SEU   simple extrapolation, unless
SSRI   selective serotonin reuptake inhibitor

1

Introduction

1.1 Medical Nihilism

Modern medicine is amazing. As the heiress of thousands of years of study into the causes and cures of disease, it can mitigate much suffering. Patients place great trust in its products, practitioners, and institutions. Fortunes are spent on research, regulation, and consumption of its primary goods—pharmaceuticals. Despite such successes and the confidence that many place in them, in this book I defend *medical nihilism*. Medical nihilism is the view that we should have little confidence in the effectiveness of medical interventions.

At first glance medical nihilism is unreasonable. Modern medicine has plenty of marvelous interventions—miracles, really—from simple drugs like antibiotics to sophisticated surgeries like coronary bypass. Moreover, views akin to medical nihilism are often adopted by people promoting dangerous movements, such as anti-vaccine campaigns, or implausible alternatives, such as homeopathy. Finally, if true, medical nihilism would render many of our quotidian practices—faithful consumption of costly medications, financial commitments to health insurance plans, research on new drugs—rather odd. So, medical nihilism seems, at first taste, a strange pill to swallow. In spite of all this, I argue here that it is the right view to take about modern medicine.

Think of ailments for which we have no cures: mysterious pains such as fibromyalgia and arthritis, many deadly forms of cancer, many neurological diseases such as Parkinson's, nearly all psychiatric diseases such as depression and bipolar disorder, and even the simplest and most ubiquitous of diseases such as the common cold. Or think of medical interventions which are widely consumed but are barely effective and have many harmful effects, such as selective serotonin reuptake inhibitors, statins, and drugs for type 2 diabetes. My arguments in this book use these as illustrative examples, but my focus is on the methodology of medical research.

The best methods that clinical scientists employ to test medical interventions—including the randomized controlled trial and meta-analysis, often said to be the pinnacle of research methods in medicine—are, in practice, not nearly as good as they are often made out to be. As a result, we ought to be skeptical of much of the evidence generated by such methods. Since such evidence is often employed to bolster the case that medical interventions are effective, doubting the veracity of such evidence contributes to undermining the case that medical interventions are effective. That said, much evidence in medical science is compelling—such evidence comes from the very best randomized controlled trials and systematic reviews (usually performed by academics who are independent of the manufacturers of the medical interventions in question), and is informed by fundamental scientific research—and very often such evidence indicates that our most commonly used medicines are essentially ineffective.

One might wonder why so much confidence is placed in medical interventions, and how it could be that such confidence—held by highly trained and conscientious physicians and their well-informed patients, and implemented by cost-concerned healthcare systems—is so often misguided. My arguments for medical nihilism go some way toward addressing this. The explanation I offer here is that the methods that are employed to test medical interventions are 'malleable': the design, execution, analysis, interpretation, publication, and marketing of medical studies involve numerous fine-grained choices, and such choices are open to being made in a variety of ways, and these decisions influence what is taken to be the pertinent evidence regarding a medical intervention under investigation. Another line of explanation, which is equally compelling and complements my focus on the methods of medical research, appeals to the fantastic financial incentives in place for selling medical interventions that seem effective. Such incentives entail that when evidence can be bent in one direction or another because of the malleability of methods, such bending is very often toward favoring medical interventions, and away from truth. Good work has been done by journalists and physicians exposing the influence of financial incentives on medical science. However, less work has been done which shows *how* it is that the methods of medical science are malleable enough to allow such influences. As a philosopher of science, this is my focus: I argue that the methods of medical research are malleable, and such malleability contributes to exaggerated claims of effectiveness of medical interventions.

Medical nihilism is not merely a tough skepticism espousing low confidence about this or that particular medical intervention. Rather, medical nihilism is a more general stance. It may be true that medical interventions ought to be assessed empirically on a case-by-case basis. However, such assessments ought to be construed broadly to include the frequency of failed medical interventions, the extent of misleading and discordant evidence in medical research, the sketchy theoretical framework on which many medical interventions are based, and the malleability of even the very best empirical methods employed to warrant causal hypotheses in medicine. If we attend more critically and broadly to our evidence, malleable methods, and background theories, and reason with our best inductive framework, then our confidence in the effectiveness of medical interventions ought to be low.

I employ the tools of philosophy of science to defend focused theses in each chapter, which, when taken together, support the overall claim of the book. The chapters in Part I articulate theoretical and conceptual groundwork. The chapters in Part II critically examine detailed nuts and bolts of the methods of medical research. The chapters in

Part III summarize the arguments for medical nihilism and what this entails for medical research and practice.

This book applies philosophical tools to scientific research to defend a radical position about medicine. But this book is a work in philosophy of science in a second (and secondary) sense, by applying scientific tools and findings to philosophical topics. In Chapter 2, for example, I defend a hybrid theory of health and disease. In Chapter 7 I criticize the view known as epistemic uniqueness, which holds that evidence uniquely justifies a particular belief. There is an emerging view in philosophy of science that in many aspects of science, facts and values are inextricably intertwined. The arguments in this book support this: in Part I the theory of disease that I elaborate requires both an empirical and a normative component, in Part II I show that social values can permeate inductive reasoning, and in Part III I note the role of values in reconsidering research, practice, and regulation in light of medical nihilism. Throughout the book I illustrate the insights that can be gained by modeling scientific inference with formal philosophical tools.

More generally, this book is an exercise in *contextualized demarcation*. Philosophers of science have long attempted to demarcate good science from bad science, or pseudo-science. A lofty ambition of philosophers has been to articulate a principle that can demarcate good science from bad in a general and context-free manner. The trouble with this ambition is that every such principle has faced counterexamples. Popper's principle of falsificationism is perhaps the most famous of these approaches. Good science, according to Popper, involves the development of theories that make precise predictions, and good scientific activity involves rigorous attempts to refute such predictions and thus the corresponding theories (Popper, 1959 [1935]). Kuhn's notion of paradigms was also influential: normal science is based on a paradigm, and empirical fact-gathering in the absence of a paradigm is not science, according to Kuhn (1962). Recently, Hoyningen-Huene (2013) argues that a distinctive feature of science is its systematicity, though critics have noted that systematicity is neither necessary nor sufficient for an endeavor to be scientific. Based on what appears to be an intellectual standstill, many philosophers have given up on the attempt to develop general, context-free principles of demarcation. But from this standstill it does not follow that one cannot identify unreliable science in a particular domain or context. Such contextualized demarcation is a central ambition of this book.

This book, then, contributes to contemporary debates in philosophy of science by focusing on technical details within science. However, the primary aim of this book is to defend a radical position about medicine by using the tools of philosophy of science. To use Godfrey-Smith's term, this book is a work in 'philosophy of nature'—assessing science, not in its raw form but rather through the lens of philosophy, to determine the most compelling view about what its message is (2009). Similarly, to use Chang's term, this book is a work in 'complementary science'—the critical reconsideration of ideas taken for granted in science, with the aim of enhancing our understanding of nature, and perhaps improving our lives (2004). This is a work of philosophy of nature or complementary science, in which the subject is one of our most important institutions, one of our best-funded sciences, and a practice upon which the most fragile and vulnerable members of our society depend.

To evaluate medical nihilism with care, toward the end of the book I state the argument in formal terms. A compelling case can be made that, on average, we ought to assign a low prior probability to a hypothesis that a medical intervention will be effective; that when presented with evidence for the hypothesis we ought to have a low estimation of the likelihood of that evidence; and similarly, that we ought to have a high prior probability of that evidence. By applying Bayes' Theorem, it follows that *even when presented with evidence for a hypothesis regarding the effectiveness of a medical intervention, we ought to have low confidence in that hypothesis.* In short, we ought to be medical nihilists. The master argument is valid, because it simply takes the form of a deductive theorem. But is it sound? The bulk of this book argues for the premises by drawing on a wide range of conceptual, methodological, empirical, and social considerations.

My use of the Bayesian apparatus is meant to provide clarity to the master argument of the book and unity to its chapter-level arguments. This tool is often employed by philosophers, but since some of my audience may not be familiar with it, I provide a simple overview of it in Appendix 1. This overview will also be helpful to students who have not yet been exposed to this elegant tool. However, this book is far from being a work in formal philosophy. I use the formalizations only as needed, and I place most of the formal work in appendices. Moreover, for the most part I avoid taking a stand on the debate about the fundamental nature of probability and the foundation of statistics, except insofar as the master argument assumes that the prior probability of a hypothesis ought to be taken into account, to the extent that it can, when assessing its posterior plausibility (even some card-carrying frequentists grant this). The arguments I raise for medical nihilism are compelling regardless of one's allegiance to a particular school of statistics or theory of scientific inference. Bayesianism is beset with well-known problems—I use it here only to unify and clarify an otherwise disparate and complex argument. This book has more in common with feminist philosophy of science, in that it is a critical examination of the methods and results of scientific practice, with attention to the social nexus in which such activity occurs.[1]

My position should not be interpreted as antithetical to evidence-based medicine, or as supportive of other critical views of medicine. On the contrary. The medical interventions that we should trust are those, and only those, that are warranted by rigorous science. The difficulty is determining exactly what this appeal to rigorous science amounts to, which is why medical nihilism is a subject for philosophy of science. Views similar to medical nihilism are shared by a cluster of critical perspectives on medicine, such as those of the antipsychiatry movement, religious opposition to particular medical practices, and the holistic and alternative medicine movements.

---

[1] Such as (Longino, 1990), (Solomon, 2001), and (Douglas, 2009).

I do not align myself with these views. Taking a critical position toward medicine does not imply an alignment with other positions critical of medicine. Indeed, most of my arguments apply to many of these movements more strongly than they do to medicine itself.

Moreover, some of the critical arguments I raise here have been made by reflective epidemiologists and physicians. The work written by physicians, epidemiologists, and science journalists supporting medical nihilism is vast.[2] These thinkers are not cranky outsiders, but rather are among the most prominent and respected physicians and epidemiologists in the world. For instance, the former editor of one of the top medical journals has claimed that "only a handful of truly important drugs have been brought to market in recent years" while the majority are "drugs of dubious benefit" (Angell, 2004b). Or consider the position of the epidemiologist John Ioannidis, suggested by the title of his important article: "Why Most Published Research Findings Are False" (2005b). The current editor of another eminent medical journal recently had this to say about contemporary medical science: "Afflicted by studies with small sample sizes, tiny effects, invalid exploratory analyses, and flagrant conflicts of interest, together with an obsession for pursuing fashionable trends of dubious importance, science has taken a turn towards darkness" (Horton, 2015). Even a leading textbook of pharmacology discusses "the inadequacies of current medicines" (Dutta, 2009). In this book I show that these are well-grounded statements of medical nihilism.

I anticipate several objections. Surely my view does not warrant distrust in all of medicine? There is no place I would rather be after a serious accident than in an intensive care unit. For a headache, aspirin; for many infections, antibiotics; for some diabetics, insulin—there are a handful of truly amazing medical interventions, many discovered between seventy and ninety years ago. However, by most measures of medical consumption—number of patients, number of dollars, number of prescriptions—the most commonly employed interventions, especially those introduced in recent decades, provide compelling warrant for medical nihilism. In the final chapters I articulate medical nihilism in response to the most salient objections.

Perhaps the most vociferous objection will be that medical nihilism is a general and empirical thesis, but I have merely selected a handful of examples that appear to be favorable to the thesis, and even in those cases the empirical evidence does not straightforwardly support medical nihilism, but does so only from an excessively pessimistic perspective. I agree that if I were offering an inductive argument based on the examples in the chapters that follow, I would be offering a weak argument indeed. To make such an inductive argument for any particular medical intervention or class of interventions would require great empirical detail, and perhaps a journalist's knack for exposition. But such is not *my* argument. I have tried to learn from such detailed

---

[2] Recent examples include books by Marcia Angell (2004b), Moynihan and Cassels (2005), Carl Elliott (2010), Ben Goldacre (2012), and Peter Gøtzsche (2013), and articles by epidemiologists such as John Ioannidis, Lisa Bero, Peter Jüni, and Jan Vandenbroucke (cited throughout this book).

arguments offered by others for a wide range of medical interventions, including critical exposés of antipsychotics, antidepressants, statins, blood pressure lowering drugs, and drugs for diabetes.[3] However, most of my conclusions are based not merely on appeals to empirical examples, but are based on principled arguments about the malleability of medical research methods, the thin theoretical basis undergirding many interventions, and the many biases and fraud in medical research. These conclusions apply far more broadly than the limited examples I employ to illustrate them. Of course, I do not deny that I have chosen my examples carefully. By focusing on the most widely prescribed classes of interventions, I hope to show that medical nihilism is a compelling view regarding the most used medical interventions today, which, given their widespread use, is troubling enough.

Nihilism has several connotations. The primary one is existential: the denial that some particular kind of value, abstract good, or form of meaning in life exists. Existential nihilism can be a metaphysical thesis, which holds that an alleged entity or other aspect of the universe does not in fact exist (this is sometimes referred to as 'antirealism' about the entity in question, and of course there are specific terms for metaphysical nihilism regarding particular entities, such as 'atheism'). Existential nihilism can be an epistemological thesis, which holds that knowledge of an aspect of the universe is impossible (this is sometimes referred to as 'skepticism'). Existential nihilism can be a justificatory thesis, which holds that widely held beliefs about an aspect of the universe are not justified based on our available evidence. The other main sort of connotation of nihilism is emotional: the feeling of despair that might be associated with any of the forms of existential denial. My use of the term should elicit various connotations in different parts of the book. There is historical precedence in applying the term 'nihilism' to medicine. The term 'nihilism' was first introduced to philosophy in the nineteenth century, and within a few decades the term 'therapeutic nihilism' was used to describe the views of prominent physicians such as Holmes, who held that the drugs of his day should be "sunk to the bottom of the sea" (1860). I discuss this historical precedence in §1.3.

Nihilism became a prominent view more generally in nineteenth-century western thought. A primary philosophical project of Friedrich Nietzsche, at the end of the nineteenth century, was to diagnose and ultimately cure such nihilism (Reginster, 2006). As above, nihilism is the view that particular important ideas or objects or values lack objective standing or cannot be realized. This is, fundamentally, the thesis argued for here with respect to medical interventions. Arguably, the highest value of medicine is the elimination of symptoms of disease and ideally the elimination of disease itself and thereby the achievement of health. Medical nihilism holds that this value very often cannot be realized and confidence in medical interventions typically lacks objective standing. Nietzsche's proposed strategy for overcoming nihilism was to

---

[3] See (Moncrieff, 2013), (Kirsch, 2011), (Moynihan & Cassels, 2005), and (Gøtzsche, 2013).

re-evaluate those values that cannot be realized. The final chapter of this book attempts to do this for medical research and treatment.

Medicine is a vast domain, with many aims and many kinds of means to attain those ends. I use the term 'medical nihilism' to align myself with nineteenth-century therapeutic nihilism—the thesis defended in this book is centrally about therapeutic interventions. But the thesis is more specific still: it is focused on the most widely used kind of therapeutic interventions in recent generations, namely, pharmaceuticals. For the most part I do not discuss surgical interventions or diagnostic instruments or other sorts of tools used in medicine (though occasionally I draw on such examples). The historian of medicine Roy Porter noted that "the prominence of medicine has lain only in small measure in its ability to make the sick well. This was always true, and remains so today" (1999). That is, medicine has been (and continues to be) an important social institution for reasons other than its ability to heal. However, a central goal of medicine is to cure diseases or at least to provide care for the diseased, and the most prominent class of interventions used to achieve this end today is pharmaceuticals. This is the subject of *Medical Nihilism*. That said, some of the arguments for medical nihilism, especially those of Part II, are focused on the methods employed in medical research— the considerations of those chapters apply to medicine generally and not just to concerns about pharmaceuticals.

Three kinds of readers should find this book valuable. It is primarily a work of philosophy of science, and so is relevant to professional philosophers and students of philosophy. The overall thesis is, of course, a view that pertains to medical research and practice, and so I hope medical scientists, physicians, funding agencies, medical students, and policy institutions will consider my thesis compelling. Finally, our society has become a voracious consumer of medicine, and thus the thesis of medical nihilism ought to concern anyone who contributes to this appetite. This book pertains to philosophers, physicians, and patients.

The risk of such a target audience is that this book may spread itself too thin on philosophical substance while being too thick for non-specialists. However, many physicians and epidemiologists are concerned with foundational problems in their discipline, and I expect this book to pique that interest and thereby motivate a critical study of the challenges facing medical research. Patients are deeply concerned about using medical interventions that work and that do not cause much harm. With the widespread availability of medical information today, patients are more informed than ever before. This book provides a critical foundation for the *use* of such information. And though my focus is on details of medical research, this is foremost a work in philosophy of science. In short, this book should appeal to many both within and outside professional philosophy.

The significance of this book is, I hope, obvious. In the central chapters I articulate the shortcomings of medical research methods, and in the final chapters I emphasize the practical implications for clinical practice, medical research, and policy.

1.2 Our Present Confidence

Since medical nihilism holds that we ought to have little confidence in medical interventions, it would help to show that our confidence in medical interventions is indeed high. Our general confidence in the effectiveness of medical interventions can be gauged in a variety of ways. Among the most frequently prescribed and best-selling classes of pharmaceuticals today are statins for lowering cholesterol; various kinds of drugs for lowering blood pressure; drugs for controlling the blood sugar of type 2 diabetics; and selective serotonin reuptake inhibitors for depression (such lists are readily available on the internet). These classes of drugs are prescribed tens of millions of times per year.

Horwitz and Wakefield (2007) note that in the United States 10 percent of women and 4 percent of men use antidepressants in any given month. People who were treated for depression were 4.5 times more likely to take a psychopharmaceutical intervention in 1997 than in 1987, and during the 1990s spending on antidepressants increased by 600 percent in the United States. The amount spent on the most widely employed classes of pharmaceuticals doubled in a recent five-year period (Moynihan & Cassels, 2005). From 1990 to 2000 the use of antidepressants among young Australians went up by a factor of ten, and the use of cholesterol-lowering drugs by Canadians went up by 300 percent. Drugs for lowering cholesterol generate over $25 billion per year in revenue. Such phenomena have led the pharmaceutical industry to be among the most profitable of all industries, year after year.

Part of the explanation for such rapid increase in the use of pharmaceuticals is a general confidence in their effectiveness—and such confidence appears to be increasing. Shannon Brownlee describes a survey in which respondents were asked if they agree that modern medicine "can cure almost any illness for people who have access to the most advanced technologies and treatments" (2008). Such unbridled optimism was expressed by over one third of respondents. This is especially striking given how few cures modern medicine has. Even one of medicine's most fantastic achievements—insulin as a treatment for type 1 diabetes (an example that I employ throughout this book)—is not a cure for this terrible disease. There are very few cures in medicine, and yet a large proportion of people think that almost any illness can be cured by available interventions.

Such confidence is especially noteworthy when the medical interventions are novel. Patients and their prescribing doctors have a fondness for novel medications. Of the ten top-selling drugs in Britain, eight were introduced in the preceding decade, and of the fifty top-selling drugs in Britain, only *two* had been on the market for more than twenty years. This is striking, since novel medications have not been tested in practice for very long—they have not withstood the test of time. Yet it is these medications that we spend the most money on. This is in part due to the fact that novel medical interventions are protected by patent, and so manufacturers can charge what they wish for the medicine without competition (a subject I discuss in Chapter 12). But as also represented by sheer numbers of prescriptions, our enthusiasm for novel medications is high.

The surging confidence in medical interventions is in part explained by the clever and tenacious marketing tactics employed by pharmaceutical companies. In 1976 Merck's outgoing chief executive expressed the aim of making drugs for healthy people, so that they could "sell to everyone" (cited in (Moynihan & Cassels, 2005)). Tens of thousands of drug company representatives engage with physicians on a regular basis in an attempt to increase physicians' confidence in the medical interventions manufactured by their employers. The vast majority of medical research today is funded by private industry, and these industrial sponsors of research control how evidence is generated and interpreted and publicized—the standard procedure is to use the results from research in whatever ways are most conducive to increase confidence in the effectiveness of medical interventions.

Another potential explanation for the apparent confidence we have in medical interventions is that most people do not understand basic aspects of experimental design, and thus do not understand the shortcomings of medical research. The National Science Foundation surveys US residents regarding their understanding of scientific concepts. One of the survey questions is a simple but fundamental question about science. There are two parts to the question:

(1) Two scientists want to know if a certain drug is effective against high blood pressure. The first scientist wants to give the drug to 1,000 people with high blood pressure and see how many of them experience lower blood pressure levels. The second scientist wants to give the drug to 500 people with high blood pressure and not give the drug to another 500 people with high blood pressure, and see how many in both groups experience lower blood pressure levels. Which is the better way to test this drug? (2) Why is it better to test the drug this way?

Sadly, in 2012 only one third of people answered the two parts correctly (NSF, 2014). (In Chapter 10 I discuss the rationale for the main features of experimental design in medical research, including the feature probed in this question.) Lacking understanding of what makes some experiments plausible (and others implausible), people could ascribe a higher degree of credibility to medical research than is warranted. I hope that this book will contribute to literacy regarding research methods in medicine.

In short, by many measures, we (patients, physicians, policy-makers) have high confidence in the effectiveness of medical interventions. But this was not always so. Medical nihilism has, until recently, been a widely held view.

1.3 Brief History of Medical Nihilism

Medical nihilism is an old view, expressed by pillars of western culture for millennia. Artists, playwrights, and essayists have been especially critical of medical treatments, though physicians themselves have been among the most outspoken medical nihilists. In what follows I give a brief historical survey of prominent proclamations of medical nihilism. Three common themes run through these historical passages as arguments for medical nihilism: the untreatability of many diseases, the ineffectiveness of many medical interventions, and the corrupting influence of money in medicine. In the chapters that follow these arguments for medical nihilism from centuries past reappear in modern guise today.

The ancient philosopher Heraclitus claimed that doctors tortured the sick and were "just as bad as the diseases they claimed to cure."[4] The Bible describes a woman who "had suffered many things of many physicians, and had spent all she had and was nothing bettered but rather grew worse" (Mark 5:25–7). Virgil's *Aeneid* (*c*.25 BC) contains the line *aegrescitque medendo*, which has been used to express the idea that "medicine increases the disease" or "the disease worsens with treatment." Hippocrates, the symbolic parent of western medicine, recognized how ineffective much of medicine in his day was, and is attributed with the saying "to do nothing is also a good remedy."

Proclamations of medical nihilism were ubiquitous in early modern Europe. Shakespeare's plays are full of jabs at physicians and their 'physic' (medicines). For instance, when Macbeth asks a doctor if he can cure a diseased mind, a troubled brain, or a heart weighted with sorrow, the doctor claims that "the patient must minister to himself" (*c*.1606). Unimpressed, Macbeth retorts, "throw physic to the dogs; I'll none of it." Shakespeare's medical nihilism was bolder in *Timon of Athens* (*c*.1605): "Trust not the physician; His antidotes are poison, and he slays." The doctor is one who gives "bitter pills" (*The Two Gentlemen of Verona*, *c*.1590), but "death will seize the doctor too" (*Cymbeline*, 1611).

Shakespeare's contemporaries agreed. The dramatists Beaumont and Fletcher wrote that "medicine [is] worse than the malady" (*Love's Cure*, 1612). Thomas Dekker, in *The Honest Whore* (*c*.1605), held that it is safer to fight a duel than to see a doctor. Ben Jonson, in *Volpone* (1606), wrote that doctors are more dangerous than the diseases they treat: doctors "flay a man, before they kill him." Another seventeenth-century British writer, Matthew Prior, wrote the sardonic line "cured yesterday of my disease, I died last night of my physician." Such medical nihilism was expressed not only by playwrights—here, for instance, is Francis Bacon: "Medicine is a science which hath been (as we have said) more professed than labored, and yet more labored than advanced: the labour having been, in my judgment, rather in circle than in progression. For I find much iteration, but small addition" (1605).

Nearly a century later writers were no more impressed with medicine. Here is John Dryden at the end of the seventeenth century (from *Epistle to John Dryden of Chesterton*), expressing a version of medical nihilism based on a theory of diseases that held that in principle it was not possible to artificially mitigate diseases (two centuries later the physician Jacob Bigelow would express the same idea with his 'self-limited' theory of disease):

---

[4] Cited by Wootton (2006), who provides many historical passages through the millennia that express medical nihilism.

> Better to hunt in fields for health unbought,
> Than fee the doctor for a nauseous draught.
> The wise for cure on exercise depend;
> God never made his work for man to mend

In addition to this principled form of medical nihilism, Dryden was also deeply suspicious of medical practice. Here is another passage (from *To John Dryden, Esq*):

> So liv'd our sires, ere doctors learn'd to kill,
> And multiplied with theirs the weekly bill.

We see in Dryden all three themes noted above as arguments for medical nihilism found throughout these historical passages: an in-principle futility of treatment ("God never made his work for man to mend"), the ineffectiveness of medications (a "nauseous draught" with which "doctors learn'd to kill"), and a corrupting financial incentive ("the weekly bill").

Early modern suspicion of medicine was not limited to the British. Descartes claimed that "medicine as it is currently practiced contains little of much use" (1637)—though Descartes expected that if medicine were to be based on modern scientific principles it would become more useful. In several of his essays Montaigne expressed medical nihilism. Voltaire, too, had little good to say about medicine, and also expressed the view that would later be called the self-limited theory of disease: "the art of medicine consists of amusing the patient while nature cures the disease." David Wootton's book (2006)—aptly subtitled 'Doctors Doing Harm since Hippocrates'—shows a lithograph by artist Honoré Daumier printed in 1833 which has a doctor brooding over one of his dead patients in a coffin, asking himself, "Why the devil do all my patients go off like this...I do my best by bleeding them, purging them, drugging them...I just don't understand it!" Several of Molière's plays are satires on medicine; every mention of a medical intervention is for Molière an opportunity for sarcasm and dark humor.

The nineteenth century saw medical nihilism become a mainstream view within medicine.[5] In a talk at the Massachusetts Medical Society in 1860, Oliver Wendell Holmes, Senior, who was Dean of the Harvard Medical School, wrote "if the whole materia medica, as now used, could be sunk to the bottom of the sea, it would be all the better for mankind—and all the worse for the fishes." A few decades earlier another eminent physician and Harvard professor, Jacob Bigelow, argued that most diseases are 'self-limited'—the course of most diseases is determined by their nature and this course cannot be influenced by medical interventions (1835). Both Holmes and Bigelow excluded palliation from their medical nihilism. Holmes did not wish opium to be sunk to the bottom of the sea ("the Creator himself seemed to prescribe" it), and

---

[5] Wootton notes, "'Therapeutic nihilism', the belief that most conventional medical therapies did not work, became the norm amongst sophisticated (particularly Parisian) doctors in the 1840s." Though the term 'therapeutic nihilism' has occasionally been used, 'medical nihilism' more precisely connotes my thesis.

Bigelow noted that one does not need medical interventions to alleviate suffering: "he who turns a pillow, or administers a seasonable draught of water to a patient, palliates his suffering." This illustrates what in Chapter 12 I call *gentle medicine.*

Nineteenth-century medical nihilism was also expressed by scholars outside the medical community. Nietzsche, who suffered from many ailments throughout his short life, wrote that "Popular medicine and popular morality belong together and ought not to be evaluated so differently as they still are: both are the most dangerous pseudo-sciences" (*Daybreak*, 1881). Tolstoy expressed such skepticism more elliptically: "Though the doctors treated him, let his blood, and gave him medications to drink, he nevertheless recovered" (*War and Peace*, 1869).

After the confidence about medical progress that attended the great success of anti-biotics and insulin from the 1920s to the 1950s, the 1960s and 1970s saw a resurgence of medical nihilism. Indeed, medical nihilism flourished in the 1970s, aligned with more general sentiments opposed to technological development, institutional author-ity, and capitalism. For instance, Illich (1975) argued for a thesis like medical nihilism and was hugely influential both within and outside medicine and academia—Illich claimed that "modern medicine is a negation of health...it makes more people sick than it heals." Similarly, McKeown (1976b) drew on historical records to argue that the increase in life spans and concomitant rise in populations was not due to public health technologies such as vaccinations, quarantines, and drugs, but rather to increases in nutrition thanks to an increase in socio-economic equality.

Discussing Illich's argument, Michel Foucault noted that much of the empirical evi-dence that Illich cited that purported to demonstrate the ineffectiveness and harmful-ness of medical interventions could be explained by errors caused by the ignorance of physicians (say, misdiagnosis), or by other kinds of human errors (say, accidental over-doses), or by institutional problems (say, administrative factors that contribute to overtreatment), rather than by problems with effectiveness of medical interventions themselves (2004 [1974]). The former kinds of problems Foucault referred to simply as iatrogenic, while the latter kind of problem, which Foucault held to be more interest-ing, he referred to as 'positive iatrogenicity': "the harmful effects of medication due not to errors of diagnosis or the accidental ingestion of those substances," but rather due to the effects of the substances themselves. This distinction between basic iatrogenicity and positive iatrogenicity is important: contemporary medical nihilism (unlike some of the historical versions of medical nihilism articulated above) is a thesis about medical interventions and the research methods employed to study those interventions, and is not about the capability of physicians.

It is one thing to argue that medicine had little ability to make the sick well in, say, the seventeenth century. It is an entirely different thing to argue that this is the case today. A predictable response to these past articulations of medical nihilism is to hold that such skeptical views were warranted in these historical contexts, but the problems that once motivated medical nihilism have been now resolved. Above I noted that there have been numerous articulations of aspects of medical nihilism in the last decade, many of which come from prominent physicians and epidemiologists. One way the today-is-different response could counter these recent articulations of medical nihilism is to claim that we now have a proper application of the scientific method to test medical interventions, there are many more effective medical interventions available to us, and these medical interventions are guaranteed to be safe and effective by diligent government regulation. The rise of the evidence-based medicine movement and the fantastic resources devoted to biomedical science—this response continues—have rendered medical nihilism tenable only as a thesis about the history of medicine. The central arguments of this book show why such a response is not compelling. In §1.5 I briefly survey these arguments. But first, it will help to understand the basic contours of contemporary evidence-based medicine.

1.4 Evidence-Based Medicine

The randomized controlled trial (RCT) became an important method for testing the effectiveness of medical interventions in the second half of the twentieth century, and is often said to be the gold standard in medical research. The first RCTs were performed in the early part of the twentieth century to test hypotheses about psychic phenomena (Hacking, 1988), but soon they were applied to the medical context, and are now held to be absolutely central to what has come to be called evidence-based medicine.

The fundamental design of a randomized trial is simple: among the subjects recruited for a study, some are allocated to receive the experimental treatment while others are allocated to receive a placebo or a competitor treatment. The crucial feature of RCTs is that this allocation of subjects is randomized: the allocation of any particular subject to a group is determined randomly. Randomized subject allocation is meant to ensure that the subjects between the two groups are, on average, similar with respect to properties that might influence the outcomes that are measured in the trial. Measurements of parameters of interest are made, usually both before the interventions are given, during the trial, and then again after. The mean values of these parameters for each group are then compared, and if the value of a parameter is different between the groups then an inference might be made that the intervention has a causal capacity to change the value of that parameter. Several chapters throughout this book, especially those of Part II, are concerned with some of the fine-grained details of RCTs. One of my central conclusions is that despite the great care that goes into the design and implementation of trials, these methods are nevertheless *malleable*: the design, implementation, analysis, interpretation, and publication of trials require numerous choices—medical researchers have wide latitude in implementing the details of their methods.

Randomized trials play a crucial role in regulation. For a new drug to be approved by the United States Food and Drug Administration (FDA), generally there must be two RCTs that suggest that the drug is superior to a placebo. One of the problems with this standard is that the FDA does not consider how many trials were performed on a drug beyond these two positive trials (in Chapters 8 and 12 I critically examine these regulatory standards for drug approval). Often, positive studies are published while negative ones are not, which creates a systematically skewed public record of the effectiveness of the drug. This phenomenon is referred to as publication bias—the ubiquity of publication bias supports my arguments in several chapters.

Contemporary evidence-based medicine (EBM) has emphasized the importance of randomized trials in testing the effectiveness of medical interventions. Indeed, medical scientists often hold that randomized trials are the best method for testing medical interventions, and that non-randomized study designs should not be trusted. Researchers who have articulated the principles of EBM have developed 'evidence hierarchies', which are rank orderings of methods based on their presumed freedom from systematic error. Randomized trials are normally at the top or close to the top of these evidence hierarchies. This has invited a great deal of criticism in recent years, especially from philosophers of science. In Chapter 5 and elsewhere I articulate some of these criticisms of evidence hierarchies, and in the remainder of Part II I identify other problems with methods in medical research that have thus far received less critical attention.

In the last several decades, meta-analysis has emerged as a common method to test the effectiveness of medical interventions. A meta-analysis is a quantitative summary of the results of multiple randomized trials or other kinds of studies. Many now hold that meta-analysis is by its very nature a better method than individual trials, and indeed is the ultimate standard for testing medical interventions. For this reason, EBM often places meta-analysis at the top of its evidence hierarchies. In Chapter 6 I argue that meta-analysis is not as good a method as many have made it out to be, and like individual trials, meta-analyses are malleable.

Although I disagree with some of the EBM principles of evidence, the ambition of my arguments offered in Part II is consistent with the central motivation of EBM—namely, to raise the standards of evidence that medical interventions must be held to when assessing their effectiveness. As was the case with the EBM movement in its early days, a result of holding medical interventions to a higher epistemic standard is that fewer medical interventions are deemed effective. A close examination of the methods of medical research supports the general thesis of medical nihilism.

1.5 The Key Arguments

1.5.1 Targeting diseases with magic bullets

To be effective, a medical intervention must improve one's health by targeting a disease. These concepts—health, disease, and effectiveness—are controversial. I lay the foundation for the remainder of the book in Part I by giving an account of what disease is and what a medical intervention must do to be deemed effective. Among the leading accounts of disease—naturalism, normativism, hybridism, and eliminativism—in Chapter 2 I defend a version of hybridism, and in Chapter 3 I further develop this account of disease and apply it to contemporary problems of disease attribution, including medicalization and overdiagnosis. A hybrid account of disease holds that for a state to be a disease that state must both have a constitutive causal basis and cause harm. The two requirements of hybridism entail that a medical intervention, to be deemed effective, must target either the constitutive causal basis of a disease or the harms caused by the disease (or ideally both). This provides a theoretical underpinning to the two principal aims of medical treatment: care and cure. If an intervention does not modulate either the constitutive causal basis of a disease or the harms caused by the disease then it cannot be effective. There are several classes of medical interventions that fall short of this standard, including interventions for conditions which are arguably not genuine diseases (such as most alleged cases of 'female sexual dysfunction' and other conditions which critics argue have been inappropriately medicalized), many interventions for pre-disease states, and some interventions which have tiny absolute effect sizes (such as antidepressants).

Chapter 2 is the most purely philosophical of the chapters. As the book proceeds my focus turns more to research methodology, and toward the end it becomes more empirical. I anticipate some readers will prefer the philosophical analysis while other readers will be impatient for hard evidence. This book has both, but the emphasis differs in different parts.

An ideal that medical interventions have been held to in the last century is based on the model of 'magic bullets'. Effective medical interventions—such as penicillin and insulin—are well characterized by this metaphor. The magic bullet model of medical interventions represents two principles: specificity and effectiveness. The magic bullet model gained currency in the mid-twentieth century with the introduction of antibiotics and insulin. Drawing on the science of pharmacodynamics, I argue in Chapter 4 that once we appreciate the complexity of the physiological basis of many diseases, and the cascading complexity of how exogenous interventions interact with our physiology, the expectation of effectiveness ought to be mitigated, and concomitantly, we ought to expect many 'side effects' (I put this Orwellian term in quotes because I prefer the term 'harm profile' to refer to unintended harmful effects of interventions—I discuss this in Chapter 9). The expectation that drugs can intervene on one or few microphysiological targets and thereby bring about an effect which is both clinically significant and symptomatically specific is, for many of our medical interventions, unfounded. The magic bullet model has been a guiding ideal for medical interventions, and the low effectiveness of many interventions can be understood in virtue of the fact that these interventions and their target diseases do not satisfy the principles of the magic bullet model. The overall thesis of this book suggests that we should place less emphasis on the magic bullet model and develop other kinds of interventions for improving health.

1.5.2 Malleability of contemporary research methods

The chapters in Part II involve fine-grained analyses of the methods used to test medical interventions. In a word, these chapters show that such methods are *malleable*. Despite the alleged merits of randomized trials and meta-analyses, I show that there are many fine-grained decisions that go into their design, evaluation, interpretation, and publication. The result is that one can *bend* these methods in a variety of ways, and such bending is usually in the direction of overestimating the effectiveness of medical interventions and underestimating their harms.

In Chapter 6 I show that different meta-analyses of the same evidence can reach contradictory conclusions. Meta-analysis fails to provide objective grounds for assessing the effectiveness of medical interventions, because numerous decisions must be made when performing a meta-analysis which allow wide latitude for subjective idiosyncrasies to influence its outcome.

For example, medical scientists employ 'quality assessment tools' to measure the quality of evidence from medical studies, especially randomized trials. These tools are designed to take into account various methodological details of medical studies, including randomization, subject allocation concealment (sometimes called 'blinding'), and other features of studies relevant to minimizing bias. There are now dozens of these tools available. In Chapter 7 I show that quality assessment tools differ widely from each other, and second-order empirical studies show that they have low inter-rater reliability and low inter-tool reliability. This is an instance of a more general problem I call the underdetermination of evidential significance. Disagreements about the strength of a particular piece of evidence can be due to different—but in principle equally good—weightings of the fine-grained methodological features. The malleability of medical research runs deep.

Measuring the effectiveness of medical interventions faces three epistemological challenges: the choice of good measuring instruments, the use of appropriate analytic measures, and the use of a reliable method of extrapolating measures from an experimental setting to a more general and less controlled setting. I argue in Chapter 8 that in practice each of these challenges contributes to overestimating the effectiveness of medical interventions. These challenges suggest corrective principles. The instruments employed in clinical research should measure patient-relevant and disease-specific parameters. Effectiveness always should be measured and reported in 'absolute' terms (using measures such as 'risk difference'), and caution should be employed when interpreting measures in 'relative' terms (such as 'relative risk reduction'). I show that employment of relative measures promotes the base rate fallacy, which leads to exaggerated claims of effectiveness. This is a serious problem because relative measures are widely reported. Finally, extrapolating from research settings to clinical settings should more rigorously take into account ways in which interventions can fail to be as effective in target populations as they are in experimental populations.

Unfortunately these principles are often not satisfied, and the result is a systematic overestimation of the effectiveness of medical interventions.

1.5.3 Harms, bias and fraud

In contrast to the systematic overestimation of the benefits of medical interventions, in Chapter 9 I argue that *harms* of medical interventions are systematically *underestimated*. Numerous factors—conceptual, methodological, and social—contribute to this underestimation. I articulate the depth of such underestimation by describing these factors. The way harms are defined and measured in research contributes to their underestimation. The basis of assessing the harm profile of drugs begins with what are called 'phase 1' trials, which involve the administration of an experimental drug in humans for the first time. Unfortunately, results from the vast majority of phase 1 trials remain unpublished. Randomized trials, despite their usual place at the top of evidence hierarchies, are another significant factor which leads to the underestimation of harms. The 'power' of a clinical trial is the ability of a trial to detect a difference of a certain effect size between the experimental group and the control group. Power is normally thought to be pertinent to detecting benefits of medical interventions. It is important, though, to distinguish between the ability of a trial to detect benefits and the ability of a trial to detect harms. I show that there are several aspects of trials which tend to maximize the former by sacrificing the latter. Furthermore, at every stage of medical research the hunt for harms is shrouded in secrecy, which contributes to the underestimation of harm profiles of medical interventions.

In Chapter 10 I give an overview of the many pervasive forms of bias in medical research. Some of these biases are exacerbated by the fantastic financial incentives and conflicts of interest present in medical research and regulation. Fraud is an extreme result of these incentives. The most pernicious form of bias in medical research today—publication bias—is properly characterized as fraud, since it very likely involves intentional exaggeration of the effectiveness of interventions by systematically distorting the published research record. Such cases are ubiquitous. The pervasiveness of bias and fraud, I argue, supports one of the premises of the master argument for medical nihilism. Given the pervasiveness of bias and the fantastic financial incentives now in place for showing that an intervention is seemingly effective, the bending of malleable research methods is almost always in favor of overestimating effectiveness and underestimating harms of medical interventions.

My arguments in the above chapters involve a close examination of the most widely employed methods in medical research. The conclusion of this line of investigation is a pronounced skepticism about the results of much medical research. This does not entail an outright dismissal of all empirical findings in medicine, however. There are better and worse token studies in medical research. The better ones tend to be performed by academics who are independent of industry, funded by non-industrial sources (say, governmental agencies), and employ better methodological safeguards. As noted above, one of the prominent problems in medical research today is publication bias, and the best meta-analyses are those that are able to overcome this by accessing all unpublished data. Throughout the book I draw on empirical findings, and I focus on the evidence from more reliable studies. Moreover, I draw on what I call second-order research, which is the empirical study of empirical studies. For example, there are meta-analyses about meta-analyses which show that industry-funded meta-analyses are vastly more likely to suggest that a drug is effective compared to those performed by independent academic researchers. Such second-order research has become more common in recent years, and results from such research support my concern about the malleability of (first-order, lower-quality) medical research.

1.6 The Master Argument

The disparate arguments described above come together to support the master argument for medical nihilism. To provide coherence to the particular chapter-level arguments and unity to the overall thesis of the book, I formulate the master argument in terms of Bayes' Theorem. In short, suppose H is a hypothesis claiming that a medical intervention is effective, and E is evidence supporting that hypothesis. The conclusions of the particular chapter-level arguments are that on average we ought to have a low $P(H)$, that we ought to have a low $P(E|H)$, and that we ought to have a high $P(E)$—I explain this in Chapter 11. Since our confidence in the effectiveness of a medical intervention, given some piece of evidence, can be represented as the conditional probability $P(H|E)$, and since, by Bayes' Theorem, this is equivalent to $P(E|H)P(H)/P(E)$, when taken together as premises in this master argument the conclusions of the chapter-level arguments entail that our confidence in the effectiveness of a medical intervention ought to be low.

In addition to drawing on the arguments from previous chapters, in Chapter 11 I introduce three widespread empirical phenomena to motivate the master argument. The first is the ubiquity of medical interventions that have been rejected because they were found to be ineffective or harmful. Such rejected medical interventions are easy to find in the history of medicine, but are also common in recent decades. The second widespread empirical finding that motivates medical nihilism is that for many of our most widely used medical interventions, the best evidence available indicates that such interventions are barely effective, if at all. The third phenomenon that motivates medical nihilism is the ubiquity of discordant evidence: for many medical interventions, some evidence suggests that they are effective while other evidence suggests that they are not. Given my arguments in Part II, one must approach empirical results in medical research with care—in Chapter 11 I rely on some of the very best randomized controlled trials, meta-analyses, and systematic reviews. These empirical findings, together with the principled arguments about medical science from earlier chapters, warrant medical nihilism.

This radical view about medical interventions invites spirited counterarguments. I end Chapter 11 by considering some predictable objections.

1.7   After Nihilism

Medical nihilism suggests modifications to ways in which certain kinds of diseases are treated. Conservative approaches to treatment have been popular throughout history. Hippocratic texts from the fifth century BC discuss many non-drug therapies, and place confidence in *vis medicatrix naturae* (the healing power of nature). Numerous proclamations of non-interventionism were made—by artists, authors, scientists, and even physicians—from the fifteenth to the nineteenth century. The twentieth century, in contrast, saw a radically aggressive approach to medical treatment become the norm. Even alternatives to mainstream medicine today involve forms of intervention (chiropractic treatment, herbal medicine, acupuncture, etc.). I co-opt a term sometimes used for this latter set of practices—*la médecine douce*, or gentle medicine—I merely borrow the term, and not their principles or practices. Gentle medicine involves therapeutic humility and is a moderate form of non-interventionism. Given the arguments raised throughout the book for medical nihilism, I argue that medicine today should be less aggressive. The arguments for medical nihilism are compelling for the most widely employed medical interventions, and for their associated ailments we could use more gentle medicine.

The arguments articulated in support of medical nihilism, especially the ones from Part II in which the malleability of research methods is emphasized, suggest modifications to ways in which medical research is carried out. There are many proposed solutions to problems raised in this book, ranging from minor modifications to medical research (such as requiring the registration of trials prior to data collection, and open access to trial data), to revolutionary changes (such as the complete socialization of medical research). In Chapter 12 I describe some of these proposals for realigning medical research which are consistent with medical nihilism. These include stricter standards for detecting benefits and harms of medical interventions, a closer scrutiny of corporate research, and a shift in the research agenda away from "drugs of dubious benefit" toward projects which are consistent with gentle medicine and which have the potential for greater impact—such as research on the importance of diet and exercise, and on neglected tropical diseases. In articulating the thesis of medical nihilism, my hope is that humanity might benefit from rethinking the art and science of medicine.

PART I
Concepts

2

Effectiveness of Medical Interventions

2.1 Effectiveness and Disease

Medicine aims to mitigate death and disease. An effective medical intervention is one that improves the health of patients by curing disease or at least treating the symptoms of disease. Effectiveness of medical interventions is a capacity to satisfy one or both of these ends. Though fine as a starting point, an analysis of effectiveness based on this platitude leaves many conceptual and practical problems unilluminated.

Some interventions are effective for minimizing shyness, or mitigating male baldness, or modulating female reproductive cycles. Other interventions have been used to intervene on homosexuality or drapetomania (a slave's urge to escape his master). At least some of these interventions are not properly 'medical,' since they are not targeting genuine diseases with the aim of improving a person's health. It is just a sociological accident, such reasoning would go, that physicians have sometimes administered such interventions. This thought, though, depends on a particular view of the appropriate aim of medicine. Restating the platitude that effective medical interventions improve health by targeting diseases does little to help distinguish effective medical interventions (say, insulin for type 1 diabetes) from medical interventions that are not effective (say, bloodletting for tuberculosis), or from interventions that are not medical (say, giving lunches to poor schoolchildren), or from interventions that do not target genuine diseases (say, cognitive behavioral therapy for homosexuality). That is because our platitude depends on the notoriously controversial notion of disease.

In what follows I canvass some of the leading accounts of disease, and defend a hybrid account, which holds that there is both a constitutive causal basis of disease and a normative basis of disease. This entails conceptual requirements for effectiveness. To be effective, I argue, a medical intervention must successfully target one or ideally both of these bases of disease. There are goals in medicine other than the treatment of disease, and interventions employed to achieve those goals—say, screening modalities, vaccinations, and methods of birth control—do not fall under the purview of my analysis, because my focus is on therapeutic interventions that are used for treating diseases with the end of improving health.

A widely held view is that health is a naturalistic notion, construed as normal biological functioning, and disease is simply departure from such normal functioning.

Alternatively, many hold a normative conception of health and disease, which claims that health is a state that we value and disease is simply a state that we disvalue. A third approach is a hybrid view, which holds that a disease has both a biological component and a normative evaluation of that biological component. A fourth major approach is eliminative, which claims that the general notion of disease should be replaced by physiological or psychological state descriptions and evaluations of such descriptions. I will call these, respectively, naturalism, normativism, hybridism, and eliminativism. A rich literature has formulated numerous considerations for and against these accounts of disease. In what follows I highlight the central issues dividing these approaches, show that these different conceptions of disease have different implications for determining what counts as an effective medical intervention, and ultimately defend hybridism and a corresponding theory of effectiveness.

To illustrate the importance of the concept of disease for understanding effectiveness, consider antidepressants, a class of interventions widely prescribed to treat depression. If, as some argue, most cases of depression are normal responses to the many difficulties of life and do not involve a departure from normal biological functioning (call these quotidian cases), then quotidian cases of depression are not cases of disease according to naturalism or hybridism.[1] It follows that for a quotidian case of depression, antidepressants cannot be considered effective, since they are not intervening on an abnormal biological function to render it normal. This point is conceptual, not empirical. The notion of effectiveness of medical interventions is not merely effectiveness simpliciter—effectiveness of medical interventions does not merely refer to a capacity for generating some effect or other. Rather, the notion refers to a capacity to improve health by modulating the constitutive causal basis or symptoms of disease.

There happen to be many empirical studies that show that antidepressants are ineffective for most cases of depression, where 'ineffective' means 'does not modify patient scores on depression measurement scales compared with patients receiving placebo.'[2] The conceptual conclusion of the line of reasoning above is that regardless of such empirical evidence, given a certain theory of disease, antidepressants cannot be effective in quotidian cases of depression, because the right way of construing 'effective' is roughly 'intervenes on constitutive causal bases or symptoms of disease to improve health,' and because quotidian cases of depression are not cases of disease.[3]

---

[1] For this interpretation of depression, see (Horwitz & Wakefield, 2007). I discuss this view in more detail below.

[2] The usual scale employed in such research is the Hamilton Rating Scale for Depression. I use this scale to illustrate problems of measurement in Chapters 8 and 9. One of the most careful reviews of the effectiveness of antidepressants concludes that positive effects of such interventions are "nonexistent to negligible among depressed patients with mild, moderate, and even severe baseline symptoms" (Fournier et al., 2010). Such findings are now ubiquitous; as examples, see (Kirsch, Moore, Scoboria, & Nicholls, 2002), (Nemeroff et al., 2003), (Ioannidis, 2008a), and (Kirsch et al., 2008). I return to such findings in Parts II and III.

[3] It would not necessarily follow that antidepressants should not be used in quotidian cases of depression—perhaps antidepressants in quotidian cases could be considered similar to coffee or wine (pleasant perks in a day of a hard but otherwise normal life)—but the use of antidepressants in quotidian cases (by this line of reasoning) would not be based on their *effectiveness*. This line of reasoning would require

One's commitment to a particular concept of disease is crucial for assessing the effectiveness of medical interventions. In Chapter 8 I address methodological concerns regarding how effectiveness is measured, and in Chapters 3 and 4 I provide more specific requirements for what medical interventions must do in order to be deemed effective. Here I am concerned with the more general conceptual question of what effectiveness is.

I defend a hybrid theory of disease. One aspect of my defense of hybridism is to argue that alleged alternatives (naturalism, normativism, and eliminativism) are not compelling. Hybridism about disease entails that for a medical intervention to be deemed effective it must successfully target either the constitutive causal basis of a disease or the harms caused by the disease (or both). Thus, the conception of disease articulated here provides a standard of effectiveness with which to assess medical interventions. Moreover, this conception of disease provides a theoretical underpinning to the two central aims of medical treatment: disease cure and symptom care.

2.2 Naturalism

Here is a prominent formulation of naturalism by Boorse (1977):

(1) The *reference class* is a natural class of organisms of uniform functional design; specifically, an age group of a sex of a species.

(2) A *normal function* of a part or process within members of the reference class is a statistically typical contribution by it to their individual survival and reproduction.

(3) *Health* in a member of the reference class is *normal functional ability*: the readiness of each internal part to perform all its normal functions on typical occasions with at least typical efficiency.

(4) A *disease* is a type of internal state that impairs health, i.e., reduces one or more functional abilities below typical efficiency.

This is a naturalist account of disease because disease is construed solely in terms of a departure from health, which itself is construed as typical biological functioning. This account has a clear implication for the notion of effectiveness of medical interventions: to be effective, according to this naturalist account, an intervention must modify an internal state that is not functioning normally and modify the functioning of the relevant part or process to bring it to typical efficiency.

Boorse's theory of disease requires a diminished ability of parts or processes to contribute to survival or reproduction for a state to count as a disease (made explicit in condition 2 above). Disease, then, involves a failure of a system to achieve its adaptive function. In contrast, some argue that a naturalist theory of health and disease is better

antidepressants to have at least some capacity to decrease patients' severity of depression, which, as the empirical work cited in footnote 2 suggests, is doubtful (see also Chapter 8).

based on a causal or mechanistic account of function.[4] An entity or activity is properly functioning, on a mechanistic account, if and only if it makes its typical contribution to the operation of the mechanistic system that contains that entity or activity. Boorse (1977) himself employs mechanistic language when he calls a disease a "failure of parts of the body to perform biological functions which it is statistically normal for them to perform," but the ultimate biological function according to Boorse is the propensity of a part to contribute to survival and reproduction. One can relax this requirement on the notion of biological function: the internal states that constitute diseases can be thought of in terms of parts of the body that perform certain operations; when these operations are not typically efficient for the end of that particular mechanism, the internal state is a disease.[5] Thus health is the capacity of one's physiological mechanisms to operate at typical efficiency; a disease is the failure of certain mechanisms to perform their particular functions at typical efficiency. I call this condition for a disease concept CAUSAL BASIS OF DISEASE.[6] The corollary condition for the concept of effectiveness—that a medical intervention must modulate the physiological basis of disease—I will call CAUSAL TARGET OF EFFECTIVENESS. This is a standard widely held among medical scientists and clinicians.[7]

The physiological properties referred to by CAUSAL BASIS OF DISEASE are distinct from etiological causes of disease. They are the states that are *constitutive* of a disease; a disease's *basis*; the pathophysiological causes of patient-level symptoms. These physiological states may have distal causes, often external to one's body (viruses, poisons, or animal fats, for example). CAUSAL BASIS OF DISEASE is about the causal constitution of a disease, and not the causal etiology of a disease. I argue below that CAUSAL BASIS OF DISEASE is a necessary but insufficient condition for a state to be considered a disease.

Consider one of my favorite examples: in 1921 Frederick Banting and his young colleague Charles Best discovered that type 1 diabetes is constituted by an inability to produce insulin. Since the function of insulin is to control blood sugar, type 1 diabetics have abnormally high blood sugar, which itself causes the phenomenological symptoms of diabetes, such as frequent urination, increased hunger, weight loss, seizures, fatigue, and eventually death. Banting and Best were able to isolate and purify insulin from laboratory animals and inject it into diabetic patients, which is a fantastically effective intervention. Diabetes is roughly characterized by a naturalist theory of disease—on either of the above notions of function, it satisfies CAUSAL BASIS OF DISEASE—and the treatment of diabetes with insulin satisfies CAUSAL TARGET OF EFFECTIVENESS. This example is prototypical: regardless of what damages the pancreas

---

[4] See (Schaffner, 1993) and (Murphy, 2008).

[5] To use the definition of Bechtel and Abrahamsen (2005), a mechanism is a "structure performing a function in virtue of its component parts, component operations, and their organization." See also (Hausman, 2012) for what he calls the 'functional efficiency theory' of health.

[6] I refer to candidate conditions for concepts of disease and corollary concepts of effectiveness of medical interventions by the script used here.

[7] The way that medical scientists have conceived of the precise nature of the constitutive causal basis of diseases has changed over time, from a monocausal theory of disease to a multicausal theory of disease.

(the distal, etiological causes of type 1 diabetes, which happen to be poorly understood), the constitutive causal basis of type 1 diabetes is the physiological state characterized by the inability to produce insulin as a result of damage to the pancreas. A fantastic intervention for this disease involves targeting its constitutive causal basis by administering exogenous insulin.

There are several classes of objections to naturalist accounts of health and disease. One is that it fails to track the way that conditions have been historically classified as disease. A stock example is that of homosexuality, which was long considered a disease, yet now it is not. This change, critics note, was not due to progress in knowledge of biological function or knowledge of the constitutive causal basis of the condition, but rather was due to a change in societal values. To such a line of criticism, though, a naturalist has a straightforward rejoinder: naturalism is based on a conceptual analysis of disease, rather than a historically accurate description of the way particular conditions were in fact categorized as disease.[8] Naturalism shows precisely what was wrong with ever thinking that homosexuality is a disease (namely, that homosexuality does not involve a reduction of biological function below typical efficiency).[9]

A more pressing problem for naturalism, and one that is pertinent for an analysis of effectiveness of medical interventions, is the determination of the reference class within which one ought to assess normality. A person's relevant biological functioning could be compared with the relevant biological functioning of all other people, or all people of the same sex, or all people of the same age category (the breadth of which would have to be determined), or all people of the same sex and age category, or all people who have experienced similar external stressors, and so on. Boorse states that the appropriate reference class is an age group of a sex in a species. But Cooper (2002) argues that appropriate reference classes may have to be finer-grained than this.[10] A person's biological functioning might be within the normal range in some reference classes but be outside the normal range in other reference classes. Determining the appropriate reference class may involve appealing to non-biological considerations of normality.

To illustrate this difficulty, consider again depression. There is some evidence, albeit inconsistent, which suggests that there are differences in biological functioning between those people diagnosed with depression and those people not diagnosed with

[8] See (Lemoine, 2013) for a discussion of the limitations of conceptual analysis regarding health and disease.

[9] However, according to Boorse's account, homosexuality is in fact considered a disease, because it interferes with reproduction. Boorse noted that because his theory of disease is non-normative, the fact that homosexuality is a disease according to his theory does not entail that it is a bad state or that it should be treated. Regardless, naturalism is not committed to an evolutionary account of function. And certainly according to hybridism (see §2.4), homosexuality is not a disease.

[10] A version of this problem has been raised by Kingma (2007). A related concern is the 'problem of common diseases': if many or most members of a population have a particular dysfunction (say, tooth decay), then an approach to determining proper functioning that is based solely on statistical features of a population will fail to consider that dysfunction as a disease. See (Millikan, 1989), (Neander, 1991), and (Schwartz, 2007).

depression, and depression and its alleged biological dysfunction are not statistically typical. Thus, if the chosen reference class is constituted by the set of all people, then a person diagnosed with depression will on average have certain abnormal biological functions. However, if the chosen reference class includes only people who have experienced similar external stressors, then it might turn out that on average a person diagnosed with depression will have statistically typical biological functioning relative to that reference class. The trouble is that there is no clear or objective way to determine the relevant reference class solely by appeal to naturalistic or biological considerations. In order to determine the relevant reference class one must appeal to considerations that are laden with value judgments that go beyond biological facts.

The difficulty of choosing a reference class for assessing normality is deep. As noted above, some hold that quotidian cases of depression are normal responses to life's difficulties, and argue that the relevant reference class for assessing the normality of one's functioning is the class of people who have experienced similar difficulties. But why this reference class? If I receive a head injury during a whiskey-soaked bar fight, is the appropriate reference class for assessing the state of my head all those people who engage in whiskey-soaked bar fights, or just all those people who enter a bar, or just all people? Of course, in any of these classes the biological facts regarding my head are the same. But the determination of the state of my head as an injury—as something abnormal—depends on the choice of reference class. In the class of people who engage in whiskey-soaked bar fights, the state of my head might be normal, and if this is the only ground upon which dysfunction is based, then this choice of reference class has an unintuitive consequence (namely, that my head is not unhealthy). Critics of naturalism would say that this is unintuitive because it is compelling to think that the state of my head *harms* me, regardless of the reference class for assessing my head's normality.[11] The head injury case is similar to the case of depression (both involve causes that mitigate health, and presumably for many quotidian cases of depression the pertinent causes are extrinsic, such as the loss of a loved one). Thus, those who hold that in quotidian cases of depression the appropriate reference class is the set of people who have experienced similar stressors must explain why the same is not true for my head injury during a whiskey-soaked bar fight. Whatever reasons those may be, they cannot be constituted by biological facts alone.

The reference class problem is closely related to the other central problem with naturalist accounts of health, and one that motivates its main competitors: determining the basis of harm.

Critics of naturalism hold that the badness of certain biological states is not determined by the biological features of those states alone. Mere departure from statistical normality is insufficient for a state to be deemed harmful. A five-foot-tall man does not have a disease merely because he has an abnormal anatomy (even if that abnormality creates

---

[11] Objecting to the analogy on the grounds that it involves an injury rather than disease would miss the point (and would rely on a thorny distinction between injury and disease).

difficulty in finding a mate to reproduce with). Not just any biological abnormality will satisfy CAUSAL BASIS OF DISEASE. Naturalist theories of health must explain why diseases are harmful, and critics claim that any basis of harm attribution will be value-laden. Boorse's theory requires a disease to involve an impairment of normal functional ability that impedes survival and reproduction. But critics have noted that evolutionary biology does not specify natural traits for populations, that humans have many goals besides those associated with survival and reproduction, and some have even claimed that evolutionary biology does not afford a distinction between normal and abnormal function.[12] Appealing instead to a mechanical account of function will not resolve this worry, since not all departures from normal mechanical functioning constitute problems for health—only those departures from normal mechanical functioning that we *care* about or that cause us *harm*, critics claim, constitute problems for health.

In short, naturalism faces several conceptual difficulties. It is noteworthy, though, that the central concerns with naturalism do not deny CAUSAL BASIS OF DISEASE as a *necessary* condition for a concept of disease. The reference class problem can be stated as: In order to determine whether or not CAUSAL BASIS OF DISEASE is satisfied, the appropriate reference class must be determined. And the normativist challenge can be stated as: In order to determine whether or not some departure from normal functioning constitutes a disease, a normative evaluation of the biological state in question is required. These challenges deny that CAUSAL BASIS OF DISEASE is *sufficient* as an explication of disease, but these challenges do not deny that this principle is *necessary*. This is not to say that no one has denied the necessity of CAUSAL BASIS OF DISEASE—I discuss such a claim in §2.4—but the most compelling challenges to naturalism do not entail a challenge to the necessity of CAUSAL BASIS OF DISEASE. In §2.4 I provide several reasons to think that CAUSAL BASIS OF DISEASE is in fact a necessary condition for the concept of disease.

We thus have the beginnings of an analysis of the conditions that a medical intervention must satisfy to be deemed effective. If CAUSAL BASIS OF DISEASE is a necessary condition for disease, then CAUSAL TARGET OF EFFECTIVENESS is a sufficient condition for effectiveness. This principle holds that a medical intervention is effective if it modulates the constitutive causal basis of a disease.

2.3 Normativism

To call a condition a disease, according to a normativist conception of disease, is to claim that a person with that condition is harmed by their condition. It is the disvalue of a condition that makes it a disease, rather than mere biological facts about the condition itself, according to normativism. To return to an example mentioned in §2.2, homosexuality was once considered a disease, but now it is not, and normativists can explain this change by noting that the change was not due to a development in our knowledge of the biological basis of sexuality, but rather was due to a development in

[12] See (Ereshefsky, 2009) and (Amundson, 2000).

society's attitude toward homosexuality—homosexuality was once broadly disvalued, and now it is not.

A normativist theory of health and disease is offered by Cooper (2002), who argues that a disease is a condition that "is a bad thing to have, that is such that we consider the afflicted person to have been unlucky, and that can potentially be medically treated." These three requirements are all necessary and jointly sufficient, according to Cooper, for a state to be considered a disease. The first is a classic normativist requirement: a state must be bad for a person. This is aligned with the view discussed in §2.2 that mere biological rareness is insufficient for a state to be a disease. Cooper's example is hair color: a redhead is biologically unusual, but does not thereby have a disease, since her hair color does not cause her harm (despite the fact that there is a constitutive causal basis of redheadedness that is statistically abnormal, namely a recessive gene that leads to high levels of the pigment pheomelanin and low levels of the pigment eumelanin). Boorse's theory also returns an intuitive verdict about hair color—assuming red hair does not lower one's propensity for survival or reproduction, then redheadedness is not a disease, even though it is statistically atypical. But if we abandon the adaptive notion of function for a causal-mechanical notion of function, the appeal to propensity for survival and reproduction is not available to distinguish rare biological states that are diseases from those that are not. The normativist has a conceptual resource that the (causal-mechanical) naturalist does not: since redheadedness is not a bad thing to have, it is not a disease.

I will call the requirement that a state be disvalued in order for it to be a disease NORMATIVE BASIS OF DISEASE.

What are the implications of NORMATIVE BASIS OF DISEASE for the notion of effectiveness of medical interventions? If NORMATIVE BASIS OF DISEASE is a necessary condition for disease, then one way for a medical intervention to be effective is if it modulates a state that is harmful and thereby mitigates the harm. I will call this NORMATIVE TARGET OF EFFECTIVENESS.

Normativism has been criticized on the grounds that it is unable to distinguish conditions that are intuitively thought of as real diseases (say, type 1 diabetes) from conditions that are alleged to be diseases merely because the conditions are disvalued. This line of criticism can seem compelling when the basis of evaluation is a value system different than one's own, though ultimately this line of criticism is misguided. For example, Ereshefsky (2009) argues against normativism by appealing to the alleged disease drapetomania. This condition, described by a physician in the southern United States in 1851, was said to be a disease that some slaves had which led them to try to escape from their masters. Ereshefsky writes:

From our contemporary perspective, we think that it is wrong to call drapetomania a disease. We believe that drapetomania was not a disease then and is not a disease now. But if you are normativist, you cannot say that those American doctors were wrong to call drapetomania a disease. All you can say is that we have different values than those nineteenth century doctors.

(2009, p. 224)

However, one can be a normativist about disease without being a relativist about values. It is not true that a normativist can *only* say that we have different values from those doctors in southern nineteenth-century United States: a normativist can say that we now have better values. Thus, a normativist could say that some doctors in the southern nineteenth-century United States were correct (narrowly construed) to call drapetomania a disease, because according to their values having slaves escape was bad, but such doctors were incorrect (broadly construed) to call drapetomania a disease, because their values upon which the badness of escaping slaves was based were unwarranted. Conditional on any set of values (including values that sanction slavery), Ereshefsky is correct to say that normativism lacks the conceptual resources to condemn the diagnosis of escaping slaves as diseased; but conditional only on *warranted* values (which do not include the sanction of slavery), it is incorrect to say that normativism lacks the conceptual resources to condemn the diagnosis of escaping slaves as diseased. A normativist could say, quite simply, that such disease attributions were based on an unjust social system (the immorality of slavery). Given some basis of warrant for values, normativism about health has the conceptual resources to criticize disease categories as appropriate or inappropriate.[13]

As suggested above, one argument in favor of normativism is that it can account for some historical vicissitudes of disease attributions. Some conditions have been considered to be a disease at one time in history, and now are no longer considered to be a disease, and an explanation for this change is that such conditions are no longer disvalued.

Someone with naturalist inclinations might say that a central problem for normativism is that it does not require biological dysfunction in order for a state to count as a disease. This objection must be articulated with caution, however. Modifying Cooper's example of the redhead, suppose that one's hair color *does* cause one harm. Suppose that in a particular society red hair is associated with sin and thus redheads are shunned. Our redhead, in this society, would be diseased according to a normativist. Though the naturalist might complain—there is nothing physically wrong with the redhead!—in fact CAUSAL BASIS OF DISEASE is satisfied under both the adaptive account of function (because the redheads are shunned they are less likely to survive and reproduce) and the causal/mechanical account of function (because the state is constituted by a rare genetic feature). Thus the normativist and the naturalist might agree on some seemingly odd disease attributions. As with drapetomania, this case can be resolved by appealing to the lack of warrant for the normative system of this society, but it cannot be resolved by appealing to biological facts alone, because there is a constitutive causal basis of the condition. Thus the normativist has a conceptual resource to doubt the attribution of redheadedness as a disease that the naturalist lacks.

---

[13] Of course, one could question where such a basis of warrant for values comes from. This is not the place to defend a general non-relativistic ethical theory.

Some harmful states, however, do not have an identifiable constitutive causal basis, and it is compelling to think (for many such states) that this is not merely an epistemic shortcoming. Think of wealth inequality, dishonesty, and a taste for country music. It is a stretch to think that there is a physiological constitution of such states. There are, of course, causes of such states, but as noted above, CAUSAL BASIS OF DISEASE is not the requirement that there be distal etiological causes of states, but rather is the require-ment that there is a constitutive causal basis of a state.[14] What can a normativist say about such states? Either a normativist can base disease attributions solely on states that are disvalued based on constitutive causal-mechanical dysfunction (as with the badness of insufficient insulin production), or else a normativist can make disease attributions on at least some states that are disvalued not because of constitutive causal-mechanical dysfunction (as with the badness of wealth inequality, dishonesty, and country music). The first option holds CAUSAL BASIS OF DISEASE to be a necessary condition of diseases, in addition to an explicit requirement that a state be harmful in order to be a disease (and thus the view would be a version of hybridism, discussed in §2.4). The second option entails that some classifications of states as diseases are unintuitive, and are better thought of as a departure from political, moral, or aesthetic values, rather than as a disease.

Take profound poverty, which is a state that any defensible value system ought to consider troubling, and anyone in that state is unlucky and is harmed in virtue of their poverty. At first pass, normativism must say that poverty is a disease. One could speak of poverty as a social disease, perhaps, but this metaphor aside, poverty is not a disease like type 1 diabetes or syphilis. The wrongness of poverty is a wrong of society, and though poverty is a harm to individuals, it is not a wrongness constitutive of an individual. There can be no penicillin for poverty.[15]

In short: either normativism collapses into hybridism or else normativism makes unintuitive disease attributions.

A way out of this dilemma is to distinguish those bad states that are diseases from those bad states that are other forms of badness, such as poverty, dishonesty, and having a taste for country music. The standard normativist way to do this is to hold that a disvalued state is a disease if society is organized such that the disvalued state is attended to by physicians, rather than, say, welfare counselors, fact checkers, or music critics. Cooper argues for a way to make such a distinction (2002). Her third condition for a state to be considered a disease is that the state must be medically treatable. Diabetes is medically treatable with insulin. Poverty is not medically treatable, though it is socially treatable. So diabetes is a disease and poverty is not.

---

[14] The distal causes of these three states are best thought of as occurring at a social level.

[15] There presently happens to be no penicillin for cancer either. However unfortunate, this is a result of the complex causal basis of most cancers (see Chapter 4). In contrast, there can be no penicillin for poverty because poverty is not constituted by a causal-mechanical dysfunction of a person (though, of course, pov-erty can be caused by such dysfunctions, as could occur when a person with a disease loses her job as a result of her disease).

A problem with this solution to the dilemma is that the *medically treatable* requirement is too expansive. For instance, I am especially drowsy in the morning, more than most, and this drowsiness is treatable with coffee, and so, together with the fact that this drowsiness is bad for me (because it impedes my work) and renders me unlucky (because my drowsiness is unusually somniferous), according to Cooper's account, my morning drowsiness is a disease. Cocaine could have an even perkier effect on my morning drowsiness. Merely from the badness and unluckiness of my morning drowsiness, and from the fact that caffeine or cocaine can treat my morning drowsiness, it should not follow that my morning drowsiness is a disease. A normativist might respond: doctors do not administer coffee or cocaine! That, however, is a thin sociological contingency upon which to base a theory of disease. Indeed, Cooper's suggestion simply pushes the question regarding disease attribution back a level. What makes interventions that mitigate diabetes *medical* while interventions that mitigate poverty or morning drowsiness non-medical? Naturalism (and hybridism) has a straightforward answer to this question, but I do not think a compelling answer can be given by normativism. Moreover, medical practitioners employ many interventions for states that are not diseases, such as contraceptive pills (Murphy, 2008). Further, physicians are unable to treat many diseases. Thus, as naturalists have long argued, the *medically treatable* condition is neither necessary nor sufficient for disease attribution.

Worse for my present purpose, given my concern with explicating the notion of effectiveness of medical interventions, is that appealing to the condition *medically treatable* is circular. This is not a problem for Cooper's account of disease itself; rather, it is a problem if one were to adopt this account of disease for the purpose of explicating effectiveness. A state is medically treatable if and only if there could be an effective medical intervention for that state. Here I am trying to determine what an effective medical intervention is; an intuitive though vacuous answer is that an effective medical intervention is one that increases health by mitigating disease; a normative theory of health requires a notion of *medically treatable*; thus, a normative theory of health requires a notion of effective medical interventions, so on this approach a notion of effective medical interventions is needed to explicate the notion of effective medical interventions. Indeed, a naturalist view is that the *medically treatable* condition gets matters backwards. A state is medically treatable *because* the state is a disease and an effective medical intervention might exist for the state, not the other way around. There is no penicillin for poverty because poverty is not a disease.

The fundamental insight of normative theories of disease is that biological dysfunction does not in itself warrant disease attributions, on any account of biological function, because disease attributions require valuations of states, and the basis of such valuations is not provided by biological facts. The problems raised in this section for normativism are challenges for NORMATIVE BASIS OF DISEASE if this condition is taken to be *sufficient* for disease attributions. But the problems are readily addressed if NORMATIVE BASIS OF DISEASE is taken to be merely a *necessary* condition for disease attribution. And I have given reasons to think that NORMATIVE BASIS OF DISEASE is indeed a necessary condition for disease attribution. Thus, one way for a medical intervention to be effective is for it to modulate a disease state that is harmful and thereby mitigate the harm caused by the disease. This condition I will call NORMATIVE TARGET OF EFFECTIVENESS.

Earlier I suggested that the criticisms of naturalism allow that CAUSAL BASIS OF DISEASE is also a necessary requirement for the concept of disease. Thus we have two proposed necessary requirements for the concept of disease, which suggests an alternative to naturalism and normativism.

2.4 Hybridism

Hybrid accounts of disease draw on insights of both naturalism and normativism. A prominent example of a hybrid account of disease in the context of psychiatry has been proposed by Wakefield.[16] Disease attribution, according to hybridism, involves two conditions: a state must be biologically dysfunctional, and that dysfunction must be harmful. Hybridism maintains both CAUSAL BASIS OF DISEASE and NORMATIVE BASIS OF DISEASE as necessary and jointly sufficient conditions for a state to be a disease.

What are the implications of a hybrid account of disease for the notion of effectiveness of medical interventions? Since hybridism holds both CAUSAL BASIS OF DISEASE and NORMATIVE BASIS OF DISEASE as necessary conditions for a state to be a disease, successfully intervening on either condition alone is sufficient for a medical intervention to be effective. That is because, since both conditions must be satisfied by a state for that state to be a disease, if an intervention modulates one of those conditions then that intervention goes at least some way toward mitigating the status of the state as a disease. A medical intervention need only satisfy one of CAUSAL TARGET OF EFFECTIVENESS or NORMATIVE TARGET OF EFFECTIVENESS to be effective.

I discussed in §2.3 the fundamental reason for thinking that NORMATIVE BASIS OF DISEASE is a necessary condition for a concept of disease, namely, that the badness of diseases cannot determined by biological facts alone. I take this to be conclusive: NORMATIVE BASIS OF DISEASE is indeed a necessary requirement for the concept of disease.

I also noted that eminent examples of diseases—such as type 1 diabetes, cancers, and infectious diseases—satisfy CAUSAL BASIS OF DISEASE. This, though, is merely suggestive, and is not a conclusive argument that it is a necessary requirement for a concept of disease. Indeed, Cooper (2002) argues against the necessity of CAUSAL BASIS OF DISEASE as follows:

Claiming that diseases must have a biological basis would be too strong because there might be some mental diseases where there is nothing wrong with the patient's brain. It might turn out, for example, that irrational phobias are completely indistinguishable from reasonable fears by the neuro-sciences.

---

[16] See (Wakefield, 1992) and (Horwitz & Wakefield, 2007).

However, indistinguishability of psychiatric states by neuroscience is not necessarily a reason to think that such states do not have causal-mechanical bases, because neuroscience may lack the technical sophistication to discern such bases. If, on the other hand, Cooper's indistinguishability claim is taken to be not merely epistemic but ontological, then, to the extent that one is committed to physicalism, one will find the claim mysterious. But there is a more practical counterargument to Cooper. A commitment to CAUSAL BASIS OF DISEASE is held by most contemporary research psychiatrists. One of the central contributors to the DSM-V, the most recent edition of the diagnostic manual of psychiatric disorders, claims that:

the implicit belief that there is an underlying, incompletely understood brain-based dysfunction for the behavioral, cognitive, emotional and physical symptom syndromes is the de facto definition of mental disorders used by most members of the DSM-5 Task Force and Work Groups   (Regier, 2012).

That is, CAUSAL BASIS OF DISEASE is assumed by leading psychiatric researchers.[17]

Moreover, CAUSAL BASIS OF DISEASE affords a critical perspective on some contemporary practices. Consider Cooper's own example, the reification of the state called 'social anxiety disorder' as a disease based on the fact that the drug paroxetine (Paxil) is alleged to have a beneficial effect on people categorized with social anxiety disorder. It is appropriate to call social anxiety disorder a disease, according to Cooper, because it is a bad thing to have, is an unlucky state, and is medically treatable. Suppose that it is true that social anxiety disorder is treatable by paroxetine. One might argue that because it is treatable by a drug, we have good reason to think that social anxiety disorder has a constitutive causal basis (even if we do not yet know what that basis is).[18]

On the other hand, suppose that one maintained the conviction that social anxiety disorder does not have a constitutive causal basis; this might be motivated by the thought that (as its name suggests) this state is constituted by phenomena at a social, rather than physiological, level. Thus if CAUSAL BASIS OF DISEASE were held as necessary for a concept of disease then social anxiety disorder would not be a disease. A social ill, perhaps, but not a disease. Accordingly, one might hold that paroxetine is not an effective medical intervention for social anxiety disorder. This would not be to deny that paroxetine has an effect on people categorized with social anxiety disorder— which is merely an empirical claim, and a modest one at that, since all sorts of things have effects on all sorts of people, such as baseball bats and coffee—but rather this is to make the conceptual point that if CAUSAL BASIS OF DISEASE is a necessary requirement for a state to be a disease, and if social anxiety disorder does not satisfy the requirement, then it is not a disease, and thus there can be no effective medical intervention

---

[17] One might wonder why the DSM continues to employ symptom-based definitions of mental disorders, if the writers of the DSM have the "implicit belief" in physical or biological ("brain-based") definitions of mental disorders.

[18] Tsou (2012) presents a detailed case to argue a similar point: pharmacological interventions have served as tools in the refinement of neurobiological theories of mental disorders.

for it. In short, CAUSAL BASIS OF DISEASE is a conceptual standard with which one can evaluate alleged interventions for an alleged disease.

Despite a commitment to CAUSAL BASIS OF DISEASE, present medicine does not understand the constitutive causal basis of many alleged diseases (in Chapter 4 I criticize 'magic bullet' discourse often used to describe the alleged mechanisms of many medications). This limitation is especially salient in psychiatry. Thus one might ask: what policy ought a hybridist have toward intervening on alleged diseases about which we lack knowledge of their constitutive causal bases, given the hybridist's commitment to CAUSAL BASIS OF DISEASE? Hybridism does not entail any particular treatment policy: not knowing a constitutive causal basis does not imply that there is no such basis, and so an epistemic limitation of not knowing the constitutive causal basis of a particular disease would not entail that the conceptual requirement of CAUSAL BASIS OF DISEASE remains unsatisfied.

In order for a medical intervention to be effective, its target must be a disease, and thus both CAUSAL BASIS OF DISEASE and NORMATIVE BASIS OF DISEASE must be satisfied. However, this is not an excessively stringent requirement, because an intervention needs to target only one of the bases of disease to be effective, thus only one of CAUSAL TARGET OF EFFECTIVENESS or NORMATIVE TARGET OF EFFECTIVENESS must be satisfied. Hybridism therefore provides an elegant theoretical underpinning to the two primary aims of medical treatment: cure and care. If a medical intervention satisfies CAUSAL TARGET OF EFFECTIVENESS then the intervention can be used to *cure* (or at least mitigate) a disease. If a medical intervention satisfies NORMATIVE TARGET OF EFFECTIVENESS then the intervention can be used to *care* for a patient with a disease. Some interventions target the constitutive causal basis of a disease—antibiotics, say— and for this reason alone are valuable, though such interventions are additionally valuable because they also eliminate the normative basis of the disease. Some interventions modulate only the symptoms of a disease without modulating the constitutive causal basis of the disease—some pain relievers, say—and for this reason alone are valuable. Some interventions modulate the constitutive causal basis of a disease's symptoms without thereby curing the disease—insulin, say—and so the constitutive causal basis of such diseases is not eliminated but the intervention provides value to the patient thanks to mitigation of symptoms (and thereby at least some mitigation of the normative basis of disease).[19] In short, hybridism explains why medicine has, as two primary aims, cure and care. Hybridism also explains why cure is more fundamental than care—because cure typically also offers care, but not vice versa.

One might hold that a way to satisfy CAUSAL TARGET OF EFFECTIVENESS could be to give a patient a drug that targets the constitutive causal basis of the patient's disease but simultaneously kills the patient due to other effects of the intervention. Detonating a nuclear bomb on a cancer patient is sufficient to eliminate the constitutive causal

---

[19] Claims of effectiveness, then, are what Alexandrova (forthcoming) calls 'mixed claims': claims that have both a factual component and a normative component.

basis of the patient's cancer, for example. However, the pre-theoretic starting point of my analysis is that an intervention is effective if (and only if) it increases a person's health by targeting a disease, and thus an intervention that decreases a person's health while targeting their disease (by killing them, say) cannot be deemed effective. (I address a recent account of what it means to increase health, by Krueger, in Chapter 3.)

Hybridism avoids the problems with naturalism and normativism noted in §2.2 and §2.3. Take the reference class problem for naturalism. Suppose a young man has a physiological abnormality—a deficiency of x, say—which causes erectile dysfunction. Further suppose that deficiency of x becomes more common as men age, until it is typical among men in their seventies. It is intuitive that the young man has a disease, while an old man who has the same physiological state that causes the same symptom does not have the disease. A hybridist could explain this in two ways, corresponding to the two requirements of hybridism. First, a hybridist could say that the difference between the old man and the young man is that the state of the former satisfies CAUSAL BASIS OF DISEASE but the state of the latter does not. To do this, the hybridist needs a way to delineate the reference classes such that the state of the young man is deemed abnormal but the state of the old man is deemed normal. Recall that naturalism has no purely 'natural' way of delineating reference classes, but a hybridist can appeal to social values such that when it comes to assessing sexual function, the appropriate reference class to assess the young man is, say, men under sixty, rather than all men.[20] Second, a hybridist could say that the difference between the old man and the young man is that the state of the former satisfies NORMATIVE BASIS OF DISEASE but the state of the latter does not. To do this, one would have to make the case that the young man is harmed by his state but the old man is not. A hybridist, again, can appeal to social values to do this.

2.5 Eliminativism

A recent alternative to the above theories of disease is that medicine will be able to do away with disease concepts altogether as we gain more knowledge regarding the pathophysiological processes underlying the states that we categorize as diseases. This view, which I call eliminativism, holds that the dispute between naturalism, normativism, and hybridism will eventually dissolve once we gain enough biological knowledge.

Depending on the precise formulation of eliminativism, it is not very different from hybridism. For instance, the form of eliminativism proposed by Ereshefsky (2009) holds that "we should frame medical discussions in terms of state descriptions and normative claims"—descriptions of physiological states or psychological states and value judgments of those states (see also (Hesslow, 1993)). Ereshefsky criticizes certain aspects of particular hybridist accounts, such as the commitment to an evolutionary

---

[20] There remains a line-drawing problem, but that is parenthetical to the reference class problem. A theory of disease cannot be burdened with solving the problem of vagueness.

account of function in Wakefield's version of hybridism, and thereby holds his own account as distinct (we have already seen, though, that hybridism does not require an evolutionary account of function). Ereshefsky argues that in technical discussions medical scientists might dispel with disease categories in favor of state descriptions (just as biologists use the term 'gene' in public forums while articulating the technical details for their colleagues without relying on the abstract term). The virtue of Ereshefsky's approach is that it affords a separation of debates regarding state descriptions from debates regarding normative evaluations of those states.

A similar proposal by Lange (2007) claims that we ought to dispel with coarse-grained disease categories in favor of finer-grained state descriptions. But nothing about hybridism is committed to coarse-grained categorizations of states as diseases. A good hybridist could say: the finer the grain of our categories, the better (this is the promise of 'personalized medicine'). If the grain happens to be at the level of biochemical or physiological mechanisms, all the better. Nevertheless, having a name for such states can still be useful—says the hybridist—for teaching medical students, communicating with patients, administering healthcare systems, and predicting patient outcomes. A state description might be "destruction of pancreatic beta cells, which causes insulin deficiency and thereby an increase in blood glucose levels," and the evaluation of such a state might be "bad," but to the medical student, patient, physician, and insurance program, the state remains type 1 diabetes.

Thus, for my purpose, the implications of eliminativism for the notion of effectiveness of medical interventions are similar to that of hybridism. The requirement of a physiological state description just is CAUSAL BASIS OF DISEASE, and the requirement of an evaluation of that state just is NORMATIVE BASIS OF DISEASE. Ereshefsky's proposal is to keep debates about state descriptions distinct from debates about their respective evaluations. That is perfectly amenable with hybridism. The key difference with respect to understanding effectiveness of medical interventions is that hybridism holds that effectiveness requires the targeting of a disease, whereas eliminativism holds that effectiveness is more simply a matter of targeting some state or other, without requiring that such a state be conceptualized as a disease.

As Cooper (2002) rightly notes, whether or not a condition is classified as a disease can have broad economic and social consequences. For example, the decision of a payer (insurance company, individual, or government) to fund treatment of a particular condition often depends on whether or not that condition is considered a disease. Replacing the employment of a disease concept with mere state descriptions would eliminate a central consideration in such decisions. Medicine is primarily concerned with intervention. One must decide which conditions to intervene on. It is often necessary that an individual patient make such a decision, but it is almost never sufficient, because other decision-makers, especially physicians and payers, must also decide that a condition should be intervened on. The conceptualization of a condition *as a disease* is often invoked to justify such decisions. For example, some interventions are considered appropriate treatments, because they target diseases, whereas other interventions are considered enhancements, because they modulate normally functioning biological processes. Granted, the distinction between treatment and enhancement is controversial, but much of the basis of this controversy depends on the concepts of health and disease. Without a disease concept one dispels with a foundation for making many of the important social, legal, and economic decisions in medicine.

In short, my preferred theory of disease is hybridism. But the difference between hybridism and eliminativism is slim. Eliminativists maintain requirements exactly like *CAUSAL BASIS OF DISEASE* and *NORMATIVE BASIS OF DISEASE* for those states deemed to be the proper targets of medical intervention.

2.6 Conclusion

I have argued that to be effective a medical intervention must meet one of two conditions: *CAUSAL TARGET OF EFFECTIVENESS* and *NORMATIVE TARGET OF EFFECTIVENESS*. In Chapter 3 I articulate several more specific requirements on effectiveness and address potential objections to this account. In Chapter 4 I describe an even more specific account of effectiveness of medical interventions, based on the 'magic bullet' model of interventions. This has been held as a regulative ideal, but I argue that few medical interventions come close to that ideal. The conclusions of Chapters 3 and 4 lend support to the master argument for medical nihilism.

3

Effectiveness and Medicalization

3.1 Introduction

What is an effective medical intervention? What are the conditions that a medical intervention must satisfy—what must a medical intervention do and what ends must it bring about—to be effective? My starting point was the platitude that an effective medical intervention improves health by targeting disease, and I argued in Chapter 2 that since the best theory of disease is hybridism, an effective medical intervention must satisfy either CAUSAL TARGET OF EFFECTIVENESS or NORMATIVE TARGET OF EFFECTIVENESS. These are abstract requirements. Here I articulate more specific requirements for the concept of effectiveness of medical interventions.

Effectiveness of medical interventions is a relational property, in which the relata are (i) a causal capacity of the intervention, and (ii) properties of a circumscribed set of people who have a particular disease. The type of causal/mechanical dysfunction that is the constitutive causal basis of a disease is a property of those people that have the disease in question, and the corresponding harm caused by the disease is a harm to those people that have the disease.[1] Since effectiveness of medical interventions is characterized by a causal capacity to intervene on the biological dysfunction of a disease or the harm that such dysfunction causes (or both), effectiveness is a relational property between (i) and (ii).

The analysis of effectiveness presented in Chapter 2 left many details about effectiveness unaddressed. A medical intervention can act at one of several physical scales or 'levels'—in §3.2 I argue that it is sometimes pragmatically useful to think of our most effective medical interventions operating at microphysiological levels (I further explore this aspect of 'magic bullets' in Chapter 4). The scope of applicability of medical interventions is another aspect of effectiveness: a medical intervention can be effective to varying degrees of generality (§3.3), though ultimately I argue that what matters for the typical patient is whether or not an intervention will be effective for that patient. The hybrid account of disease defended in Chapter 2 affords a critical stance on several troubling phenomena in medical science, including medicalization

---

[1] Of course, a token case of disease may also cause harm to people other than the person who has the disease, such as the strain caused by caring for someone with Alzheimer's. Nevertheless, for a condition to count as a disease, according to the hybrid account of disease defended in Chapter 2, the condition must cause harm to the person with the condition.

or 'disease-mongering' (§3.4) and overdiagnosis and overtreatment (§3.5). To use the terminology introduced in Chapter 2, most of my analysis in this chapter and throughout the book is focused on interventions that target CAUSAL BASIS OF DISEASE (this has been the focus of mainstream medicine for at least the last century). However, in §3.6 I note that interventions can independently target NORMATIVE BASIS OF DISEASE, and since interventions that target CAUSAL BASIS OF DISEASE motivate medical nihilism, we should be spurred to support more research on interventions that independently target NORMATIVE BASIS OF DISEASE (I discuss this in Chapter 12). In §3.7 I address several objections that might be raised against the hybrid account of effectiveness presented in Chapter 2 and its further articulation and application in this chapter.

3.2 Levels of Effectiveness

A medical intervention might have a physiological effect (it intervenes on a physiological mechanism, say), a clinical effect (it modifies objectively measurable symptoms, say), a patient-relevant effect (it modifies subjective reports of well-being, say), and a population-relevant effect (it modifies the average longevity of a population, say) (Ashcroft, 2002). One might assume that effects at higher levels supervene on effects at lower levels: population-relevant effects supervene on patient-relevant effects, patient-relevant effects supervene on clinical effects, and clinical effects supervene on physiological effects. CAUSAL TARGET OF EFFECTIVENESS seems to prioritize the importance of an intervention's effects at a microphysiological level. But why emphasize the microphysiological level? A friend of supervenience will hold that if an intervention modulates parameters at a microphysiological level, then the real causal action must be happening at a lower level still. By this line of reasoning, type 1 diabetes is not constituted by an incapacity to produce insulin by one's pancreas, but rather is constituted by the configuration of atoms that undergirds this incapacity, and the effectiveness of therapeutic insulin involves the targeting of this atomic configuration. And why stop at atoms?! Thinking about an ontology of levels for diseases and the causal action of medical interventions can lead to an intellectual quagmire. Some are suspicious even of the very idea of levels.[2]

However, for some of our most effective medical interventions—those aptly described as magic bullets (I articulate this notion in Chapter 4)—it is at least pragmatically useful to think of their effects as operating primarily at a particular level, usually microphysiological, and these effects have concomitant effects at higher functional levels at which a patient's phenomenological experience of symptoms occurs. Some antibiotics, for example, interfere with the mechanisms of cell growth of infectious bacteria, and it is useful to characterize the effectiveness of such antibiotics at this scale

---

[2] Thalos (2013), for example, argues that nature does not come packaged into neat levels. On the other hand, for a detailed account of levels as they are employed in explanations in neuroscience, see (Craver, 2007).

(of cellular machinery).[3] For our most impressive medical interventions—such as antibiotics and insulin—it is compelling to say that CAUSAL TARGET OF EFFECTIVENESS is satisfied, despite reservations about an ontology of levels.

For an intervention to be effective, it will not suffice that it has an effect at just any level (perhaps microphysiological, perhaps clinical, perhaps...). A medical intervention may have an effect on one level but not another, and that level may be unimportant with respect to both CAUSAL BASIS OF DISEASE and NORMATIVE BASIS OF DISEASE. Consider the practice of modulating cholesterol levels to avoid heart disease. Merely modulating a parameter such as cholesterol concentration in the blood is insufficient to satisfy CAUSAL TARGET OF EFFECTIVENESS when attempting to intervene on heart disease. It might not matter to a patient that he has cholesterol above a certain threshold, because this is not likely to cause any phenomenological symptoms (let alone heart disease or death), and therefore a drug that is effective only at lowering cholesterol would be ineffective at modulating parameters that matter to the patient. As critics of cholesterol-lowering drugs note, what matters is whether or not a patient develops heart disease, but medical scientists have employed cholesterol levels as measured outcomes in many trials (that is, cholesterol level has been used as a 'surrogate outcome' in trials on drugs intended to mitigate heart disease). The constitutive causal basis of heart disease is much more complicated than cholesterol levels—lowering cholesterol levels to mitigate heart disease does not satisfy CAUSAL TARGET OF EFFECTIVENESS. This explains the low effect sizes of such drugs for patient-relevant outcomes (see Chapter 8).[4]

Likewise, merely mitigating any harm is insufficient to satisfy NORMATIVE TARGET OF EFFECTIVENESS. That is because, for a medical intervention to be deemed effective, it must target a genuine disease. Consider again an example that I used in Chapter 2: suppose antidepressants do in fact improve patients' moods in quotidian cases of depression, and further suppose that such cases are not genuine diseases because they do not satisfy either CAUSAL BASIS OF DISEASE or NORMATIVE BASIS OF DISEASE—some might be tempted to call such drugs effective for such cases, but the hybridism I defended in Chapter 2 holds that interventions can only be effective if they target genuine diseases, and thus antidepressants cannot be effective for such cases.

Intervening on the constitutive causal basis of a disease is emphasized in medicine—this is what the psychiatrist R. D. Laing called the 'medical model.' In psychiatry,

---

[3] For some diseases and interventions we have some epistemic access to the relations of effects at various levels—insulin binds to a cellular receptor, causing many microphysiological effects, which decreases concentrations of sugar in the blood, which mitigates neurological and cardiovascular symptoms. But for many interventions we have limited epistemic access to their effects at various levels and relations between levels. For example, some drugs for multiple sclerosis appear to reduce 'white lesions,' purported to be 'biomarkers' or physiological correlates of the disease, but these drugs have little impact on symptoms (Lavery, Verhey, & Waldman, 2014).

[4] A review concluded that for men who have not suffered heart disease, these drugs have only "small and clinically hardly relevant improvement" (Vrecer, Turk, Drinovec, & Mrhar, 2003). See also (Moynihan & Cassels, 2005).

diagnosis departs from a strict medical model because diagnoses are based on syndromes, or clusters of symptoms, without understanding the constitutive causal bases of diseases. The psychiatrist Nassir Ghaemi argues for resuscitating a strict medical model for psychiatric nosology, on the grounds that understanding the constitutive causal basis of diseases could provide a foundation for developing effective interventions.[5] Ghaemi further argues that since we now have such a thin understanding of the constitutive causal basis of psychiatric diseases, psychiatrists should be therapeutically conservative (this supports my proposal for 'gentle medicine' in Chapter 12).

It takes more than merely modifying some physiological parameter or other to satisfy CAUSAL TARGET OF EFFECTIVENESS. This condition holds that an intervention must modulate physiological parameters associated with the particular disease being treated. Not all effects of a medical intervention are relevant to the targeted disease, however. Many effects of medical interventions are irrelevant to the disease in question or are themselves harms (side effects or adverse effects). Interventions target the constitutive causal basis of a disease to varying degrees of specificity (Chapter 4). Harmful side effects are familiar examples of effects of medical interventions that modulate parameters other than those that constitute a disease. But an effect of a medical intervention that does not modulate parameters that constitute a disease does not have to cause harm. Such an effect might modulate a parameter in a way that provides some non-disease benefit to a patient, or in a way that provides neither harm nor benefit to the patient, while not modulating parameters relevant to the disease being treated. An intervention with only such effects as these would not satisfy CAUSAL TARGET OF EFFECTIVENESS and thus would not be effective for the disease being treated.

In an insightful article Krueger provides a plausible account of degrees of health achieved by interventions (2015). Krueger argues that, according to naturalism, medical interventions aim at returning biological parts or processes to normal functioning. But, Krueger notes, many interventions do not in fact do this, because they replace or circumvent the pathological part or process. Organ transplants replace organs, torn ligaments are replaced by artificial ligaments, and many drugs bypass normal functions. These interventions improve health without restoring the parts or processes to normal function. To say this, though, requires an account of comparative judgments of health, or 'degrees' of health, in terms of closeness to normal function. This could be based on the relative proportion of parts of processes which function normally. To take this approach one would need a way to individuate and count the pathological parts and processes in an individual (and, I add, their relative significance).

Krueger gives an example to show how complex and indeterminate this could be. Consider a genetic disease based on a single mutation, and try to count pathological parts and processes. Do we count every token instance of this DNA sequence? Or do we count only those sequences that would have produced a functional protein if it were not mutated? Or should we count the number of mutated proteins? What about the

---

[5] See (Laing, 2011 [1968]) and (Ghaemi, 2012), and also (Murphy, 2006).

mutated sequences in tissues that do not require the normal protein? Without a unique individuation and count of pathological parts and processes we have contradictory assessments of degree of health. Krueger's solution is to focus on higher-order functional systems rather than parts or processes to give an account of degree of health: "the probability of functional efficiency with respect to the overall functional goals of the relevant whole system" (2015). As I argued in Chapter 2, these goals need not be based on evolutionary theory, and can be informed by normative considerations.

This has an implication for what it means to satisfy CAUSAL TARGET OF EFFECTIVE-NESS. This principle cannot simply require returning biological parts or processes to normal functioning. To consider one of my running examples: exogenous therapeutic insulin does not return damaged pancreatic beta cells to normal function. Nevertheless, insulin is an incredibly effective intervention for type 1 diabetes. It increases the health of diabetic patients, and this can be put in Krueger's terms: therapeutic insulin increases the probability of functional efficiency with respect to overall functional goals for patients with type 1 diabetes.

There is a straightforward sense in which exogenous insulin also satisfies NORMA-TIVE TARGET OF EFFECTIVENESS: the physiological effects of therapeutic insulin mitigate the diabetic's phenomenological symptoms which constitute at least some of their harms. This way of satisfying NORMATIVE TARGET OF EFFECTIVENESS fundamentally depends on the satisfaction of CAUSAL TARGET OF EFFECTIVENESS. In my discussion of levels of effectiveness in this section I have mostly focused on targeting CAUSAL BASIS OF DISEASE. However, NORMATIVE BASIS OF DISEASE can be targeted without targeting CAUSAL BASIS OF DISEASE (I develop this idea in §3.6).

3.3 Scope Requirements for Effectiveness

Medical interventions can be effective to varying degrees of demographic and contextual scope. The distinction between efficacy and effectiveness goes some way to addressing this. Efficacy is a causal property of an intervention which manifests in a particular controlled setting, whereas effectiveness is a causal property of an intervention which has the capacity to manifest in settings more general and less controlled than the particular experimental setting in which efficacy was demonstrated. Adding to the difference of context between experimental trials and real-world clinical settings, the demographic scope of these contexts is typically very different: clinical trials employ numerous exclusion and inclusion criteria to select research subjects, but once a medical intervention has been approved for clinical use such demographic constraints do not apply (physicians are typically permitted wide latitude in prescribing medical interventions). The usual way of thinking about the notion of effectiveness, as distinct from efficacy, holds that effectiveness is a more general property than mere causal efficacy which manifested in a particular experimental environment. The following condition is therefore insufficient as a scope requirement for effectiveness:

WORKS SOMEWHERE

An intervention satisfied CAUSAL TARGET OF EFFECTIVENESS or NORMATIVE TARGET OF EFFECTIVENESS for a group of subjects in a particular experimental setting.

WORKS SOMEWHERE might represent the concept of efficacy, but does not provide a compelling account of the scope requirement for effectiveness. Cartwright (2012) distinguishes WORKS SOMEWHERE from two other kinds of causal claims that are candidate scope requirements for effectiveness, and which are suggested by the usual distinction with efficacy. Since Cartwright is mostly concerned with causal claims in social policy, I modify her account to apply to medical interventions. Another candidate scope requirement for effectiveness is:

WORKS GENERALLY

An intervention satisfies CAUSAL TARGET OF EFFECTIVENESS or NORMATIVE TARGET OF EFFECTIVENESS for a class of patients in varied clinical settings.

An intervention could be effective for certain types of people in certain circumstances despite not being universally effective.[6] The demarcation of the class of people for whom the intervention is effective can be as broadly or narrowly defined as the details of the particular disease and intervention require. WORKS GENERALLY is the view that a medical intervention should work for a well-defined class of people and circumstances.

Another candidate scope requirement for effectiveness is:

WORKS FOR ME

An intervention will satisfy CAUSAL TARGET OF EFFECTIVENESS or NORMATIVE TARGET OF EFFECTIVENESS for this particular patient.

Although the causal claim in WORKS FOR ME is more specific than the causal claim in WORKS GENERALLY, the epistemological requirement for warranting a claim of the former type is stronger than the epistemological requirement for warranting a claim of the latter type. That is because claims of type WORKS FOR ME require justification for WORKS GENERALLY, plus justification that the specific conditions regarding the particular patient under consideration are such that the causal claim expressed by WORKS GENERALLY applies to this particular patient (see Chapter 8).

For WORKS GENERALLY and WORKS FOR ME it is *patients* in routine clinical practice who are intervened on, whereas the target of intervention in WORKS SOMEWHERE are research *subjects* (subjects are typically patients who have been selected in careful ways—see Chapter 8 for a discussion of inclusion and exclusion criteria used to filter the kinds of patients that end up as subjects). This reflects the idea that the causal claim

---

[6] A widespread view in evidence-based medicine is that although the epistemological standards for warranting hypotheses of the type WORKS SOMEWHERE are high, there are few additional requirements for warranting hypotheses of the type WORKS GENERALLY. In the domain of social policy this has been forcefully criticized by Cartwright (2012). For shortcomings of this assumption in medicine, see (Fuller, 2013a), and Chapters 5, 8, and 9.

in WORKS SOMEWHERE is limited to an experimental setting, whereas the causal claims in WORKS GENERALLY and WORKS FOR ME apply to clinical practice. The verb tense of the three kinds of causal claims represents the typical ways in which such claims are made—we talk of a particular RCT, in the past, as having suggested that some causal relation was instantiated; we talk of certain interventions, in the present, as having causal capacities; and we talk of certain interventions, in the future, as manifesting this capacity when one will use it.

In a brief account of effectiveness of medical interventions, Howick (2011b) adopts a principle akin to WORKS FOR ME as one of three necessary conditions. An effective medical intervention, according to Howick, must "be applicable to the patient being treated." This condition is obviously important—it is what clinicians and patients want to know anytime they decide whether or not to use a particular medical intervention. The effectiveness of a medical intervention is constituted in part by a causal capacity of the intervention, and it is reasonable to hope that this capacity transcends the idiosyncratic details of any particular patient (Ashcroft, 2002). This capacity may or may not operate for a particular patient, for mundane reasons (a drug for male potency will not manifest this capacity when administered to a female), for more subtle reasons (a birth control pill may not increase the risk of thrombosis in a particular woman, despite its capacity for doing so, because it inhibited pregnancy in that woman, which itself increases the risk of thrombosis), and because the constitutive causal basis of many diseases is so complex (a statin may not prevent a heart attack in a particular patient despite its capacity for lowering cholesterol, because the causes of heart attacks are manifold and complex). The interest of a patient is whether or not a medical intervention will manifest its capacity *for that patient*, and so it is scope requirements of the type WORKS FOR ME that matter for effectiveness.

I have spoken thus far as if CAUSAL TARGET OF EFFECTIVENESS and NORMATIVE TARGET OF EFFECTIVENESS are conditions that are either satisfied or not by some particular medical intervention. This idealization permitted me to explore aspects of effectiveness without pesky complications. Medical interventions can satisfy the conditions of effectiveness to varying degrees—effectiveness is something to be measured. I examine the measurement of effectiveness in Chapter 8, and argue that measuring effectiveness has three significant methodological challenges that contribute to overestimations of effectiveness.

3.4 An Ill for Every Pill

The hybrid account of effectiveness of medical interventions presented in Chapter 2 provides a conceptual standard with which one can evaluate problematic practices regarding disease definitions and subsequent diagnoses (and treatments) based on those definitions.

'Disease-mongering' involves the loosening of disease definitions or the outright construction of spurious diseases, thereby increasing the number of people who can be diagnosed with a disease. A related term is 'medicalization,' which involves describing normal problems (such as sadness or shyness or obesity) as diseases to be diagnosed and treated (Illich, 1975). Since disease-mongering increases the number of people who can be diagnosed as diseased, the practice can enrich healthcare providers and pharmaceutical companies.[7] There is an obvious financial incentive for disease-mongering.

This financial incentive can create conflicts of interest (Chapter 10). Consider the example of cholesterol levels. As with many conditions that are alleged to be diseases or precursors to disease, the definition of what counts as high cholesterol has been loosened over time (that is, the threshold for high cholesterol has been lowered), which has increased the number of people deemed unhealthy due to high cholesterol. The basis of this lowered threshold is controversial for clinical reasons and for the biased context in which the definition was formulated. Moynihan and Cassels (2005) report that of the nine experts on the panel that revised the guidelines regarding cholesterol levels in 2004, eight had financial ties to pharmaceutical companies that manufactured cholesterol-lowering drugs. The guidance provided by this panel increased the number of people who could be prescribed products of these companies by many millions.

The financial incentive for disease-mongering has been exacerbated by a feature of our patent system. If the holder of a patent can devise a new use for the patented product, that product can maintain its patent protection for that new use. For example, as the patent was expiring on fluoxetine (Prozac), Eli Lilly, the manufacturer, supported the description and definition of a new disease called premenstrual dysphoric disorder, a form of premenstrual syndrome. Eli Lilly rebranded fluoxetine as Sarafem (and repackaged it in a gender-normative lavender and pick capsule), got approval from the FDA for marketing Sarafem for premenstrual dysphoric disorder, and thereby got extended patent protection on fluoxetine. Critics claim that premenstrual dysphoric disorder is not a genuine disease but rather is a spurious category carved out of normal human experience constructed with the help of a pharmaceutical company to sell more of its product.

The disease-mongering strategy is to market a disease and only indirectly market their corresponding medical intervention: disease-awareness campaigns, funded by industry, focus on relatively mild but seemingly widespread conditions; well-known doctors, actors, athletes, and television personalities are paid by industry to emphasize the importance of the condition; patient-advocacy groups, funded by industry, promote the new disease category and demand action from policy-makers.[8] Getting approval

---

[7] A compelling illustration is provided by the former chief executive officer of Merck claiming that he wanted Merck to be more like a chewing gum manufacturer, to allow the company to make drugs for people without diseases and thereby "sell to everyone." Cited in (Moynihan & Cassels, 2005); the original source is (Robertson, 1976). See (Cooper, 2013) and (González-Moreno, Saborido, & Teira, 2015) for further discussion of disease-mongering.

[8] Here is Kenneth Kendler, a prominent research psychiatrist and contributor to the DSM, discussing such reification of disease categories: "social forces—reimbursement policies, grant review committees and journal editors—sometimes enforce a false hegemony for diagnostic manuals beyond that intended by its creators or warranted by the quality of its often tentative scientific support" (2012).

from the FDA for a new use of a pre-existing drug is much quicker and cheaper than getting approval for a new drug (I discuss the FDA evaluation of new drugs in Chapters 8 and 12). Just as critics decry the obsession with hunting for a pill for every ill, disease-mongering amounts to hunting for an ill for every pill. This is the cynical view of medicalization and disease-mongering.

The cynical view contrasts with the naïve view of medicalization and disease-mongering, which holds that new conditions like premenstrual dysphoric disorder are simply thus far undiscovered diseases, and some pre-existing pharmaceuticals just happen to be efficacious for the symptoms of these diseases. Disease-awareness campaigns and screening programs can identify and help those people who needlessly suffer from the disease.

Both the cynical view and the naïve view of medicalization can seem, at first glance, compelling. The hybrid account of disease I defended in Chapter 2 permits a balanced position between these two views. With the naïve view, hybridism holds that a medicalized condition may be a genuine disease, as long as both CAUSAL BASIS OF DISEASE and NORMATIVE BASIS OF DISEASE are satisfied. With the cynical view, hybridism holds that a medicalized condition may be a spurious disease, as long as one of CAUSAL BASIS OF DISEASE or NORMATIVE BASIS OF DISEASE is not satisfied. On the whole, the resources of hybridism provide some grounding for the cynical view. The cynical view can be understood as holding that medicalized conditions often do not have an identifiable constitutive causal basis or that the normative assumptions required to hold that the condition is harmful are misguided.

Cynics have many examples to illustrate their position: restless leg syndrome, female sexual dysfunction, erectile dysfunction, male balding, halitosis (bad breath), irritable male syndrome (not joking), bipolar disorder, attention deficit hyperactivity disorder, osteoporosis, and social anxiety disorder, to name a few. A classic description of disease-mongering was presented by Payer (1992). Disease-mongering tactics, according to Payer, include: describing a normal state (such as sadness) as abnormal, loosening diagnostic standards for a disease so that more people are diagnosable with the disease (and including very common conditions such as fatigue and irritability in the diagnostic criteria), offering a nebulous theory of the causal basis of the disease with little scientific support, recruiting opinion leaders to promote the disease category, and misusing statistical measures to exaggerate the effectiveness of the available intervention. These tactics are widely employed today.

The example of cholesterol levels, discussed above, shows that disease-mongering is related to the use of surrogate outcomes in clinical research, because one way to increase the number of people diagnosable with disease is to lower the standards for what counts as a disease, which is precisely what has happened with cholesterol levels. Cholesterol level is taken to be a diagnostic sign, and cholesterol levels have become the target of many drugs. Unfortunately many clinical trials only evaluate interventions for their capacity to modulate surrogate outcomes such as cholesterol levels. And the threshold for cholesterol levels deemed pathological has been decreased over time, greatly increasing the number of potential patients.[9] As argued in Chapter 2 and §3.2, interventions that merely target biological correlates or surrogates of a disease, but neither the CAUSAL BASIS OF DISEASE nor the NORMATIVE BASIS OF DISEASE, are not effective. To repeat, it is a necessary condition for a medical intervention to be effective that it satisfy either CAUSAL TARGET OF EFFECTIVENESS or NORMATIVE TARGET OF EFFECTIVENESS or both.

3.5 Overdiagnosis and Overtreatment

If normal states are medicalized and spurious disease categories are constructed, then more people will be inappropriately diagnosed with spurious diseases, and some of those people will be inappropriately treated. Medicalization and disease-mongering are closely related to two other phenomena that have recently attracted much criticism: overdiagnosis and overtreatment. Overdiagnosis occurs when a person is accurately diagnosed as having the pathophysiological basis of a disease but such pathophysiology would never have caused symptoms of that disease in that person's life. If such a person is treated for their pathophysiology, then that is a case of overtreatment.

Overdiagnosis and overtreatment are not just a result of expanded disease categories, however. Critics argue that overdiagnosis and overtreatment also result from searching too hard for diseases—what Alan Cassells calls 'seeking sickness' (2012). The harder we look for diseases, the more cases of disease we will diagnose. For some of the accurately diagnosed people their condition would not have caused them symptoms whether or not they are treated, and so were those people to be treated their treatment would be unnecessary.

Screening programs can lead to overdiagnosis and overtreatment for a number of reasons. One reason that screening leads to overdiagnosis was mentioned above: screening can find presumed cases of disease that would have gone entirely unnoticed by the patient had they not been screened. To use the terminology from Chapter 2, screening programs can find cases in which CAUSAL BASIS OF DISEASE is satisfied but NORMATIVE BASIS OF DISEASE is not. Thus, according to the hybridism defended in Chapter 2, such cases are not genuine cases of disease. These cases can sometimes appear dangerous but are in fact harmless, such as cancerous growths that would never harm the person. But because the condition appears dangerous, the patient might decide to intervene on it. Since the condition never would have harmed the patient, the patient would be overtreated.

---

[9] Trials of cholesterol-lowering drugs aimed at patients with LDL cholesterol levels above 240 mg/dL in 1986, then over 200 mg/dL in 1988, then over 130 mg/dL in 1994, then 100 mg/dL in 1995. See (González-Moreno et al., 2015). In a review of 192 published RCTs that compared statins to other statins or non-statin drugs, 189 RCTs reported data only on surrogate outcomes (Bero, Oostvogel, Bacchetti, & Lee, 2007). About 98 percent of these trials on statins did not even attempt to show that either CAUSAL TARGET OF EFFECTIVENESS or NORMATIVE TARGET OF EFFECTIVENESS were satisfied. See also Chapter 8.

One might hold the view that all forms of abnormal physiology (such as cancerous growths) are dangerous and thus should be intervened upon. However, as argued in Chapter 2, it takes more than mere abnormal physiology for a condition to be a disease. In fact many token cases of abnormal physiological states such as cancerous growths are harmless.[10] For example, autopsy studies reveal that a large proportion of men who die from other causes have non-symptomatic prostate cancer.

Another type of case, familiar to teachers of probability theory, demonstrates how screening programs can lead to high numbers of false positive diagnoses. Even for an excellent screening test—that is, for a screening test with a very high specificity and sensitivity, or in other words a very low false positive and false negative rate—if the disease being screened for is rare, then the vast majority of people who test positive for the disease will not in fact have the disease, and thus are liable to be falsely diagnosed. I give a specific quantitative example of this in Appendix 1, in which a screening test for a disease that occurs in 1 in 5000 people is 99 percent accurate: for such a test, if the test result is positive then there is nevertheless only a 2 percent chance that the person tested actually has the disease. If people in such circumstances are diagnosed with the disease in question on the basis of such a test, then there is a very high probability of false positives. And people who are falsely diagnosed are at risk of being inappropriately treated.

These problems with screening programs are not decisive objections against such programs. There have been numerous empirical evaluations of the benefits and harms of screening programs, and a robust finding is that some screening programs for some diseases can decrease mortality caused by the screened disease, but no screening program decreases all-cause mortality.[11] Cynics explain this by suggesting that screening programs might increase mortality by causes other than the diseases screened for, perhaps by side effects and stress caused by the screening program. A more modest explanation is that some screening programs might in fact cause a decrease in all-cause mortality but the decrease is too tiny to be detected by empirical evaluations of the screening programs. Regardless of the benefits of screening programs, as critics have noted for reasons I describe above, screening programs contribute to false positive diagnoses, overdiagnosis, and overtreatment.

Another general factor that contributes to overdiagnosis is the broadening of disease categories, mentioned above in my discussion of disease-mongering. Typically, for a patient to be diagnosed with a disease they must meet certain requirements—their blood pressure or cholesterol levels have to be above a certain value, for example. When those requirements become easier to satisfy—say, by lowering the threshold for a measured physical parameter to be deemed symptomatic—then the disease category is broadened, because more people will satisfy the requirements and thus be diagnosable

---

[10] Patz et al. (2014) argue, for example, that about 20 percent of lung cancers found in low-dose computed tomography scans, a very sensitive screening tool, are harmless. See also (Plutynski, 2017).

[11] See (Saquib, Saquib, & Ioannidis, 2015).

with the disease. Disease categories are in fact being broadened (Moynihan et al., 2013). If the disease category is broadened too far—say, by incorporating thresholds for measured parameters that are not markers of a genuine disease—then the broadened disease category will contribute to overdiagnosis.

Yet another class of factors that contributes to overtreatment is a result of the institutional and legal context of medical practice. Some physicians feel pressure to prescribe out of a concern that if they do not then they are liable to be sued for malpractice. Another form of institutional pressure on physicians comes from cost-cutting clinics and hospitals—if patients could spend more time in the clinic or hospital, then their physicians could devote more time to properly addressing the constitutive causal basis or normative basis of their disease, but since there is financial pressure to move patients quickly out of the clinic or hospital, physicians are often forced to employ a hasty approach to treatment, which often involves the use of excessive pharmaceuticals. Physicians, moreover, are often paid on a 'fee for service' basis, which rewards quick turnover of patients and thus incentivizes the writing of more prescriptions than they otherwise would write.[12]

A striking illustration of overtreatment involves the use of psychiatric medications in infants. In 2014 nearly 20,000 prescriptions for antipsychotic medications such as risperidone (Risperdal) and quetiapine (Seroquel) were written for infants aged two years and younger, and fluoxetine (Prozac) was prescribed 83,000 times for such infants. The journalist investigating this phenomenon interviewed a dozen experts in child psychiatry, none of whom could explain why such young children would be treated with these psychiatric medications (Schwartz, 2015).

3.6 Targeting the Normative

In Chapter 2 I argued that for an intervention to be deemed effective it can target NORMATIVE BASIS OF DISEASE, and since interventions that target CAUSAL BASIS OF DISEASE support the thesis of medical nihilism, we ought to support more research on interventions that might independently satisfy NORMATIVE TARGET OF EFFECTIVENESS. There are many such possibilities, from simple actions already widely employed to more complex social changes. Recall the nineteenth-century medical nihilist Jacob Bigelow, who noted that one does not need the sorts of interventions that are normally thought of as 'medical' to alleviate suffering: "he who turns a pillow, or administers a seasonable draught of water to a patient, palliates his suffering" (1835). This is an expression of gentle medicine, an idea I explore in Chapter 12. As noted in Chapter 1, nineteenth-century medical nihilists excluded palliative interventions like painkillers from their skepticism—one way to make sense of this is that painkillers typically do not target CAUSAL BASIS OF DISEASE but since they can mitigate the pains caused by a

---

[12] See (Welch, 2016).

disease, and pains are usually harms, painkillers can target NORMATIVE BASIS OF DISEASE.

There are much more radical ways to target NORMATIVE BASIS OF DISEASE. The spurious normative basis of spurious diseases—say, the once alleged harm of homosexuality—can be targeted by rethinking the false normative assumptions underlying the attribution of harm. The genuine normative basis of spurious diseases, such as female sexual dysfunction, can be targeted by rethinking the social structures and unmet needs that generate such harms. Social changes can also alleviate the genuine normative basis of some diseases—for example, the harm caused by the immobility of a paraplegic can be mitigated by modifying infrastructure to improve wheelchair accessibility. Since the focus of medicine in the last century has been the magic bullet model of interventions (see Chapter 4), that is my focus in this book. But since, as I argue in the following chapters, the magic bullet ideal has failed us for so many interventions and for so many diseases, we ought to think about other ways to improve health, which could involve targeting, in creative and radical ways, the normative basis of diseases.

3.7 Objections

In Chapter 2 I argued that to be effective a medical intervention must meet one of two conditions: CAUSAL TARGET OF EFFECTIVENESS and NORMATIVE TARGET OF EFFECT-IVENESS. Medical interventions widely deemed to be effective, such as insulin and penicillin, satisfy both of these conditions; medical interventions that are effective at mitigating only the symptoms of a disease satisfy the latter condition. In this section I address possible objections to the hybrid account of effectiveness offered in Chapter 2 and its further articulation and application to problems of disease attribution discussed here in Chapter 3.

Perhaps the foremost objection holds that effectiveness is merely an empirical matter. If a carefully conducted randomized trial shows that a medical intervention modifies a particular measured parameter—this objection goes—then that intervention is, by definition, effective. If another ten or a hundred trials show similar results, all the better. This objection depends on an impoverished construal of effectiveness. I have argued that effectiveness is a theoretical concept that depends upon other theoretical concepts, most importantly that of disease. The conceptual content of effectiveness entails that it is not sufficient for an intervention to modulate just any parameter, regardless of the strength of that modulation or the number of trials in which such modulation has been replicated.

One might hold that the view proposed here has too narrow a view of the goals of medicine. Physicians do more than simply treat diseases—cosmetic surgery, abortion, and guidance on contraceptive practices, for example, are all activities of physicians—and the standards of effectiveness defended here do not apply to such activities. I grant the multifaceted goals of medicine and the plural activities of physicians. One central and definitive goal of medicine, however, is the improvement of health by intervening on disease, and it is for this goal that my analysis applies.

One might note that many medical interventions now employed do not meet either of the conditions of effectiveness (as suggested by several of my examples), and thus my account does not reflect medical practice. However, the account of effectiveness of medical interventions proposed here is prescriptive, rather than a description of features of interventions alleged to be effective. To the extent that there is a mismatch between the standards of effectiveness of medical interventions that I have proposed here and what are taken to be effective medical interventions, I hope that my account is revisionary. That said, my account of effectiveness does track those medical interventions that are generally considered to be highly effective compared with those that are not. As examples: insulin and penicillin satisfy CAUSAL TARGET OF EFFECTIVENESS and NORMATIVE TARGET OF EFFECTIVENESS, while it is controversial whether or not statins and antidepressants satisfy either of these conditions. I explore this in further detail in Chapter 4.

A similar objection could be pressed: the above conditions that a medical intervention must meet in order to be deemed effective constitute too high a standard. According to this stringent standard, very few medical interventions ought to be deemed effective. This response, of course, is not an objection to the arguments that warrant the view itself, but is merely an objection to a possible entailment of the view. The conditions are in fact satisfied by the best medical interventions, like insulin for treating diabetes. Insulin, and other interventions like it that satisfy the conditions proposed here, serve as gold standards of effective medical interventions, and show that the conditions are attainable. In Chapter 8 I argue that the measurement of effectiveness faces three methodological problems, and in practice these problems contribute to overestimating effectiveness. If these problems of measurement were better addressed, our estimations of the effectiveness of many medical interventions would be more accurate and lower than they now are. This is in part due to the fact that very few medical interventions are magic bullets.

4

Magic Bullets

4.1 Introduction

Unchaste and unlucky one hundred years ago, a single sore on your skin would develop into sores and rashes in your mouth, anus, and genitals. After passing the disease to your newborn children, the French pox (as the English called it) would grossly disfigure your face and cause seizures and eventually dementia. Diagnosed with syphilis, your treatment with mercury would have been worse for you than no treatment at all.

In the first years of the twentieth century Paul Ehrlich was hunting for magic bullets—chemicals that would bind only to a particular kind of infectious organism (magic) and kill it (bullet). The cause of syphilis had been identified as the bacterium *Treponema pallidum*, and Ehrlich's colleague Sahachiro Hata had developed a rabbit model of syphilis to test experimental interventions that target this bacterium. They started with a molecule composed of arsenic surrounded by carbon rings, and they systematically modified the compound to create hundreds of variants, which they tested on their rabbits. They eventually discovered arsphenamine—patented as Salvarsan 606, the sixth organoarsenic that Ehrlich's team tested in the sixth group of compounds that they had developed—which was effective in treating syphilis.

Arsphenamine was a breakthrough in medicine, not only because it was an effective treatment of syphilis (and as a result was for years the most prescribed drug in the world), but because it was among the first highly specific interventions. Medical interventions until the early twentieth century were general in their (purported) mechanisms of action, and particular interventions were employed for a wide variety of diseases. For example, perhaps the most common intervention for two thousand years was bloodletting, employed to treat smallpox, scurvy, lunacy, fever, labor pains, and nearly every other ailment for which one might decide to see a physician. Mercury, the standard treatment for syphilis until 1910, was for centuries widely employed to treat many other ailments—as a sixteenth-century writer put it, "mercury is noble, useful in many fields."[1] As late as 1818 a Scottish doctor prescribed mercury for nearly all diseases, and another nineteenth-century doctor prescribed quinine for all two thousand diseases that he was aware of (Foucault, 1973).

---

[1] Cited in (Fleck, 1979), p. 4. Fleck used the earlier term *Spirochaeta pallida* for the causative microorganism of syphilis. See (Wootton, 2006) for a sustained discussion of pre-twentieth-century medical interventions which, though they were widely employed, were almost entirely ineffective.

Arsphenamine, on the other hand, targeted only particular disease entities, and intervened only slightly on one's normal endogenous physiological systems. Given its method of production, its chemical purity, and its specific mechanism of action, it was among the first modern chemotherapies. To describe this drug Ehrlich used the term magic bullet.

Not all magic bullets are for killing targets, as arsphenamine kills the bacteria that cause syphilis. Some magic bullets deliver important physiological products into the bodies of those who need them. One of the first and most striking examples of such a magic bullet is insulin, developed a decade after arsphenamine. In 1921 Banting and Best isolated insulin from dogs and administered the first clinical insulin to a teenage boy dying of type 1 diabetes in Toronto General Hospital. This boy's diabetic symptoms were quickly alleviated. Encouraged by this success, Banting and Best administered insulin to dozens of comatose children in a diabetic ward—the legend is that by the time they were injecting insulin into the last of the children in the ward, the first were waking from their starvation-induced comas.[2]

Another magic bullet from this era was penicillin, discovered in 1928 by Alexander Fleming. Antibiotics revolutionized medicine because they were the first effective treatments for a wide array of infectious diseases (penicillin, incidentally, replaced arsphenamine as the standard treatment for syphilis), and targeted invading bacteria typically with little influence on normal physiology.

The magic bullet model is based on two principles of medical interventions—specificity and effectiveness—and a related model of disease which holds that targets of magic bullets have a simple constitutive causal basis (§4.2). Pharmacodynamics is the study of the mechanisms of drugs and their biochemical and physiological effects. The principles of magic bullets have been important in the development of pharmacodynamics, but despite the continued use of the magic bullet model to describe medical interventions and motivate medical research, scientists who study pharmacodynamics recognize that the magic bullet model is at best a guiding ideal. There have been very few magic bullets introduced since the development of the classical magic bullets noted above (§4.3). That is because many pharmaceuticals lack the specificity of action of magic bullets (§4.4), and for many pharmaceuticals the targeted disease is characterized by a complex causal nexus (§4.5), which renders interventions less effective than they would be if the constitutive causal basis of targeted diseases was simple.

4.2 Magic Bullets

The magic bullet metaphor is powerful, and continues to serve as an exemplar for present-day medical interventions.

---

[2] Nobel laureate Peter Medawar (1983) called the discovery of insulin "the first great triumph of medical science." Medawar claimed that insulin "brought to an end, with an appropriately reverberant thunderclap, the long epoch of therapeutic nihilism." Putting aside the presumption that a single discovery could bring to an end therapeutic nihilism, the sentiment Medawar expressed illustrates how impressive insulin was (and is).

Magic bullets intervene selectively on biochemical targets, and are effective at modulating the symptoms of the targeted disease or altogether curing the disease. In terms of Chapter 2, magic bullets target CAUSAL BASIS OF DISEASE or NORMATIVE BASIS OF DISEASE (or both) and do so specifically. Most drugs elicit their effects by interacting with microphysiological entities or processes in a patient, thereby modulating those entities or processes and causing subsequent physiological changes. The term 'receptor' or 'target' refers to the microphysiological molecule with which a drug interacts. Many medical interventions act as 'ligands'—substances that bind to molecules and thereby modulate the activity of those molecules.

Such biochemical targeting can take a variety of forms. A drug can bind to enzymes, structural proteins, gene modulators, or channels in cell membranes (as examples), and such binding can increase or decrease the activity of those targets. Treating diseases often involves intervening in biochemical pathways or mechanisms—chains of chemical reactions and microphysiological activities between organic molecules, which are crucial for cell functioning.[3] These interventions can involve repairing defective pathways that are part of one's normal physiology (as insulin does for type 1 diabetes) or blocking harmful pathways that are not part of one's normal physiology (as penicillin does for infectious bacteria). Ligands (drugs) can activate or inhibit microphysiological pathways.

Penicillin, for example, binds to the bacterial enzyme transpeptidase. This enzyme produces links in bacteria cell walls as the bacteria are growing and dividing, but when penicillin is bound to the enzyme it cannot produce these links, which causes the death of the bacteria. Penicillin is effective at killing the bacteria that constitute many infections and thereby eliminating the infection and associated symptoms. Moreover, since penicillin typically does not intervene on targets other than transpeptidase, and since normal human physiology does not produce or use the targeted enzyme, penicillin is specific in its mechanism of action, and thus has few side effects.[4]

In short, the magic bullet model is characterized by two principles:

Effectiveness
An intervention is effective if and only if it successfully targets a disease and thereby improves the health of the person treated to near normality. In terms of Chapter 2, an intervention is effective if and only if it successfully mitigates CAUSAL BASIS OF DISEASE or NORMATIVE BASIS OF DISEASE (or both).

---

[3] See Thagard (2003) for a discussion of pathways. Understanding the magic bullet model is aided by thinking about mechanisms. According to Bechtel and Abrahamsen (2005), "A mechanism is a structure performing a function in virtue of its component parts, component operations, and their organization. The orchestrated functioning of the mechanism is responsible for one or more phenomena." Another oft-cited definition of mechanisms is in (Machamer, Darden, & Craver, 2000). See (Blumenthal & Garrison, 2011) for a survey of mechanisms of drug action.

[4] Of course, the development of resistance to antibiotics is a major problem, and though this is an effect of consuming antibiotics, it is not a direct harm to the patient who consumes the antibiotic, and thus is not a typical side effect. Nevertheless, the evolution of antibiotic resistance suggests an important constraint on the magic bullet model: interventions can be magic bullets for some time but later lose their potency.

Specificity
An intervention is specific if and only if it targets nothing other than the disease being treated and thus has few adverse effects. In terms of Chapter 2, an intervention is specific if and only if it mitigates nothing other than CAUSAL BASIS OF DISEASE or NORMATIVE BASIS OF DISEASE (or both).

Arsphenamine, penicillin, and insulin approach the magic bullet ideal. They are relatively specific in their mechanism of action, and they are extremely effective at intervening on diseases to improve health.[5] These two principles—*Effectiveness* and *Specificity*—are ideals, thus far stated quite vaguely. They rely on the conceptually thorny notion of disease—my account of this notion presented in Chapter 2 undergirds my analysis of magic bullets.

To articulate the magic bullet model in more detail I draw on resources from the science of pharmacodynamics. In Chapter 2 I articulated general requirements on the notion of *Effectiveness*. A related concept from pharmacology is 'potency.' Potent medical interventions cause physiological effects at lower doses than low-potent interventions. Potency depends on two properties of drugs: the ability of a ligand (the drug molecule) to bind to its targeted receptor (this is called 'affinity'), and the ability of the ligand-receptor complex to evoke a physiological response (this is called 'efficacy').[6] Potency is determined by affinity and efficacy: a medical intervention with high affinity and high efficacy has high potency. For interventions that act as ligands, some minimal degree of potency is a necessary condition to be deemed effective, though it is far from sufficient. One reason why potency is insufficient for effectiveness is the complex causal basis of many diseases (§4.5). Moreover, an intervention could be effective but not act as a ligand, in which case the notion of potency is inapplicable, and thus potency is in principle unnecessary for effectiveness (a reason for characterizing magic bullets with the more general notion of effectiveness rather than the more specific notion of potency).

The *Specificity* principle is central to the magic bullet model because if an intervention targets only the constitutive causal basis of disease, and if the physiological entities and activities that constitute the causal basis of disease are not widespread in one's normal physiology, then the intervention will likely have few side effects. In Ehrlich's own words: "We must search for magic bullets. We must strike the parasites, and the parasites only, if possible, and to do this, we must learn to aim with chemical substances!" Targeting "the parasites only" was Ehrlich's way of articulating the importance of *Specificity*, at least with respect to infectious diseases.

---

[5] Was arsphenamine really a magic bullet? Ehrlich claimed that it was; indeed, he claimed that it was the first. However, arsphenamine contains arsenic, and when improperly administered could cause tissue damage and death—though it was more specific and effective than competitor treatments for syphilis, this suggests that the magic bullet model cannot be entirely realized.

[6] Though closely related, this is not to be confused with 'clinical efficacy,' which is the capacity of a medical intervention to produce a measurable effect in the context of a trial.

'Specificity' is a term of art in pharmacodynamics. A related term of art is 'selectivity.' The difference between the two is subtle. A 'selective' ligand is one that binds to a particular kind of receptor. A 'specific' ligand is one that has a particular physiological effect. The notions are related, of course, because of the relationship between ligand-receptor binding and physiological effects. However, not all selective ligands are specific, because, as noted above and in more detail in §4.4, the relationship between ligand-receptor binding and physiological effects can be complex. Nevertheless, it is reasonable to explain the lack of specificity of a medical intervention in terms of its lack of selectivity, since non-selectivity usually implies non-specificity. Since the importance of magic bullets is their capacity to elicit health-improving physiological effects, the term 'specificity' is the more appropriate one for characterizing magic bullets. But selectivity of ligands is crucially important, and I appeal to it in the following discussion. Just as *Effectiveness* is a more general notion than potency, *Specificity* is a more general notion than the way the term is used in pharmacodynamics. A magic bullet must target only the disease being treated and not target other (normal) physiological systems, but one particular magic bullet could elicit various health-improving effects (I illustrate this with penicillin below), and thus would satisfy *Specificity* but not be specific in the pharmacodynamic sense.

The importance of *Specificity* is, I hope, obvious. In case it is not, we can reconsider a thought experiment raised in Chapter 2. One might think that a way to satisfy CAUSAL TARGET OF EFFECTIVENESS (that is, having an intervention that targets the constitutive causal basis of disease) would be to give a patient a drug that fully targets the constitutive causal basis of disease while simultaneously killing the patient. The example was detonating a nuclear bomb on a cancer patient to eliminate her cancer. The reason noted in Chapter 2 for why this intervention would not be effective was that an intervention is effective if (and only if) it increases a person's health by targeting a disease, and the nuclear bomb obviously does not increase the patient's health. Another way to make this point is that the nuclear bomb does not satisfy *Specificity*—it targets the cancer but it also targets everything else about the patient. (This thought experiment illustrates the further point that *Specificity* and *Effectiveness* can trade off against each other.)

One might think that some of the examples of magic bullets above do not satisfy *Specificity*. Consider penicillin, an iconic magic bullet. Penicillin is used to treat many kinds of infections caused by different species of bacteria. One should note that there are many different kinds of penicillin, including benzylpenicillin, phenoxymethylpenicillin, methicillin, oxacillin, and ampicillin. That aside, each of these kinds of penicillin can be effective against multiple kinds of infections of multiple species of bacteria, raising the question of whether or not penicillin satisfies *Specificity*. If penicillin does not satisfy *Specificity*, one might think that the requirement is too strong, since if anything is a magic bullet, penicillin is.

In spite of its ability to target multiple species of bacteria, penicillin does in fact satisfy the specificity requirement, because *Specificity* is about the ability of an intervention to discriminate between disease targets and normal physiological targets.

Penicillin does this extremely well. Moreover, the related notion 'selectivity' is about types of targets, not tokens of targets. As noted above, penicillin selectively binds to a particular enzyme (transpeptidase) that some kinds of bacteria use to build their cell walls. Because bacteria must constantly rebuild their cell walls, and because penicillin blocks this from occurring, penicillin is an effective intervention against many bacterial infections. Note, though, that penicillin is also selective, though such selectivity is with respect to types of targets, not tokens of targets. The various bacteria that penicillin is effective against share the same target type: transpeptidase. Penicillin targets this enzyme selectively.

That selectivity should be about types of targets rather than tokens of targets should come as no surprise, since one must think about the effectiveness of medical interventions as effectiveness against not just one particular token of a disease, but rather as against any token of that disease, generally (and thus one thinks of the effectiveness of interventions in terms of types of targets rather than tokens of targets). The constitutive causal basis of some diseases is shared with other diseases. The bacteria that cause a case of meningitis, a case of syphilis, and a case of sepsis might all be different species and cause different kinds of diseases, but all might require the activity of transpeptidase to survive. Since penicillin targets transpeptidase, penicillin is effective against multiple kinds of diseases and yet penicillin nevertheless acts selectively, since it targets only a single type of causal basis of disease. Moreover, penicillin satisfies *Specificity*, since for each disease it targets, it targets only that disease and not normal endogenous physiological systems.

Above I noted examples of outdated medical interventions that were employed for many diseases, such as quinine and mercury and bloodletting prior to the twentieth century, and suggested that these interventions are general in their mechanism of action. If a magic bullet can target multiple diseases and yet still satisfy *Specificity*, why is that not the case with interventions such as bloodletting? The answer, quite simply, is that such interventions have effects on many other physiological entities and activities besides the targeted constitutive causal basis of disease (most such interventions only target entities and activities other than the alleged constitutive causal basis of disease, and thus do not satisfy *Effectiveness* either). That an intervention allegedly targets a wide range of diseases (such as mercury before the twentieth century) is not in itself reason to think that *Specificity* is not satisfied (as indicated by the example of penicillin). But a wide range of targeted physiological entities and activities does entail that *Specificity* is not satisfied.

I will restate the two conditions. To satisfy *Specificity*, interventions must target *only* the microphysiological entities or processes that are the constitutive causal basis of the disease or the entities or processes that cause the harms of the disease (or both). To satisfy *Effectiveness*, interventions must target those microphysiological entities or processes that are the constitutive causal basis of the disease or the entities or processes that cause the harms of the disease (or both). In terms of Chapter 2, *Effectiveness* requires that CAUSAL TARGET OF EFFECTIVENESS or NORMATIVE TARGET OF EFFECTIVENESS (or both) is satisfied. To be a magic bullet, an intervention must satisfy both *Specificity* and *Effectiveness*. The magic bullet model is, of course, an idealization, and to varying degrees interventions can approximate the ideal.

Specificity and *Effectiveness* might appear to be theory-relative, or, to use a notion from Kuhn (1962), these principles might seem to be relative to a paradigm. Just like contemporary magic bullets, ancient remedies were employed with the intention of balancing physiological parameters, and the hope was that such interventions were specific (relative to the dominant physiological theory of the time). Humoral theory was, for millennia, the dominant physiological theory of health and disease—diseases were characterized as an imbalance of humors (blood, phlegm, black bile, and yellow bile), and interventions such as mercury were employed with the hope of rebalancing one's humors. Under humoral theory, interventions like bloodletting could be considered specific, since the intervention was thought to specifically target the postulated imbalance of a particular humor (blood). However, with the rise of mechanistic theories of physiology and later the germ theory of infection, specificity of medical interventions took on a new meaning, and with the benefit of our contemporary perspective, informed by modern mechanistic physiological theory, interventions such as mercury or bloodletting are rightly deemed non-specific.

The diseases treated by the classical magic bullets—bacterial infections and hormone and vitamin deficiencies—are prime examples of the monocausal model of disease. Robert Koch, one of the founders of bacteriology at the end of the nineteenth century, expressed the monocausal model of infectious diseases as follows: "each disease is caused by one particular microbe—and by one alone. Only an anthrax microbe causes anthrax; only a typhoid microbe can cause typhoid fever." This model of disease became dominant at the end of the nineteenth century, associated with the rise of germ theory and the discovery of vitamin and hormone deficiencies. Koch's model was intended for infectious diseases, but a simple causal model of diseases can be used to characterize other diseases that are targeted by magic bullets (such as diseases of deficiency). Precisely characterizing such a model would take me astray; the important point is that the constitutive causal basis of the diseases that magic bullets intervene on is *simple*. It is the simplicity of the causal basis of the diseases for which magic bullets are employed that allows for the specificity and effectiveness of magic bullets.

Characterizing the constitutive causal basis of a disease, even those that I am referring to as simple, is often not straightforward. For example, the symptoms of type 1 diabetes are caused by many physiological phenomena, and all of these phenomena are themselves caused by a lack of insulin. This lack of insulin itself has a cause, namely, the destruction of insulin-producing beta cells of the pancreas. Beta cell death itself has a causal basis, but its causes are manifold, and not well understood.[7] When used as

---

[7] A number of gene variants contribute to susceptibility to beta cell death, and a range of environmental factors have been identified as potential contributing causes of beta cell death, including diet, viruses, chemicals, and drugs.

an intervention against type 1 diabetes, in what sense is the constitutive causal basis of the target of therapeutic insulin simple? Therapeutic insulin relieves the symptoms of diabetes but does not cure the disease. Despite the manifold causes of the death of pancreatic beta cells, proximally there is only one factor that causes the symptoms of diabetes, namely, the absence of insulin. Thus, although diabetes is characterized by many causes of many symptoms, there is a single bottleneck factor that constitutes type 1 diabetes, namely, an incapacity to produce insulin. It is the corresponding dearth of endogenous insulin that is modulated by therapeutic insulin. Though therapeutic insulin does not cure diabetes, its palliative effect is great—type 1 diabetes is a disease that is otherwise torturous and fatal but is managed with therapeutic insulin. Insulin is an example of an intervention that satisfies both CAUSAL TARGET OF EFFECT-IVENESS and NORMATIVE TARGET OF EFFECTIVENESS, and does so quite specifically. Thus insulin is a good example of a magic bullet.

The magic bullet model is an idealization. For example, exogenous insulin has side effects, prominently including hypoglycemia, which can lead to cognitive impairment, abnormally rapid heart rates, and seizures. To take another of the exemplars discussed above, many people are allergic to penicillin, and widespread use of antibiotics has led to deadly antibiotic-resistant bacteria. Strictly speaking, there is no ideal magic bullet. There are interventions that can to varying degrees approximate the ideal. But there are not many.

4.3 Medicine without Magic

The classical magic bullets discussed above are regulative ideals for contemporary pharmaceuticals. Pharmacologists developing new interventions aim to satisfy the principles of magic bullets, physicians are taught to think of interventions in terms of these principles, and the popular press characterizes interventions as if they satisfied the principles of magic bullets. Unfortunately, there are very few magic bullets in contemporary medicine. Few recent medical interventions have joined arsphenamine, penicillin, and insulin as interventions that approach the magic bullet ideal. Although there have been some medical interventions introduced in recent decades that approach the magic bullet ideal, the most frequently consumed medicines today are far from being magic bullets.[8] Despite this dearth of new magic bullets, the metaphor of magic bullets is often used to describe new medical interventions, even when such interventions fall far short of the ideal principles of the magic bullet model.

The magic bullet model is often invoked, implicitly or explicitly, as an explanation for the alleged or expected effectiveness of medical interventions. Many pharmaceuticals

---

[8] Marcia Angell (former editor of *The New England Journal of Medicine*) called this the "darkest secret" of the pharmaceutical industry: "the stream of new drugs has slowed to a trickle, and few of them are innovative in any sense of that word" (2004a).

are said to selectively intervene on microphysiological targets. Here, for example, is a prominent textbook of pharmacology:

Increasing knowledge about the underlying causes of diseases is enabling the discovery of more selective and less toxic drugs. Progress in molecular biology (e.g. sequencing of the human genome, proteomics, pharmacogenomics and protein engineering) is creating new avenues for understanding precise disease mechanisms (biochemical pathways) and the discovery of new targets based on new disease pathways. (Dutta, 2009)

However, the textbook notes that although the *Effectiveness* and *Specificity* principles are ideals for medical interventions, many diseases do not have interventions that satisfy these ideals; the author gives as examples some infectious diseases (such as malaria and tuberculosis) and bone diseases (such as arthritis and osteoporosis), to which one could add most cancers, most psychiatric diseases, most chronic diseases, and most lifestyle diseases (such as obesity and addiction). The textbook notes that, despite the optimism expressed in the quote above, the complexity of the microphysiological basis of many diseases and the manifold effects of many interventions make it difficult to develop magic bullets: "the early assumption that a ligand acts at one receptor is no longer tenable and it is now well established that many endogenous ligands act at different receptor subtypes." I will return to this concern in the following sections.

One often hears of novel medications described as 'game-changers' for a particular disease, and a justification for such optimism is a description of the mechanisms of action of the intervention articulated in terms of the magic bullet model.[9] Pharmacological science, medical training, and the popular press frequently appeal to the magic bullet model when describing contemporary medical interventions. Such rhetoric can sound compelling. For example, when discussing medical interventions that are intended to enhance pathway function, Thagard (2003) describes the example of drugs for type 2 diabetes in the class thiazolidinediones, including rosiglitazone (Avandia) and pioglitazone (Actos):

These drugs reduce cells' acquired resistance to insulin by activating PPARγ, a peroxisome proliferator receptor. The peroxisome in a cell is responsible for many functions such as the breakdown of fatty acids. Stimulation of the PPARγ pathway improves the ability of cells to use insulin to incorporate glucose from the blood stream. (p. 246)

This makes it sound as if drugs like rosiglitazone satisfy *Effectiveness* and *Specificity*. Unfortunately, rosiglitazone is far from a magic bullet. I return to the example of rosiglitazone in later chapters, but in short, careful meta-analyses published since

---

[9] Such rhetoric is widespread. For example, a psychologist described fluoxetine (Prozac) as a "revolution in psychopharmacology because of its selectivity on the serotonin system; it was a drug with the precision of a Scud missile, launched miles away from its target only to land, with a proud flare, right on the enemy's roof" (Slater, 1999). Here the magic bullet model has been exaggerated by appealing to a weapon much more powerful than bullets.

Thagard made this claim show rosiglitazone to have serious harmful effects, including an increased risk of heart attacks.[10]

The names of some classes of drugs suggest that they satisfy *Effectiveness* and *Specificity*. Consider: selective estrogen receptor modulators, selective serotonin reuptake inhibitors (sometimes called serotonin-specific reuptake inhibitors), and cardioselective beta blockers. These names are misleading, because, unfortunately, most new drugs, including these classes of drugs, are not specific. The vast majority of medical interventions introduced in the last several decades have very small effect sizes and have a plethora of harmful side effects. In Chapters 9 and 11 I support this claim with many empirical examples, and below I offer some reasons for why this is so.

I am not claiming that there have been no new magic bullets in recent years. There have been some recent interventions that approach the ideal. For example, imatinib (Gleevec) inhibits the action of an enzyme (tyrosine kinase) that exists only in chronic myelogenous leukemia (CML) cells, and thus is an especially effective treatment. This form of leukemia is not like most other cancers, in that it is caused by a single malfunctioning protein, and so its constitutive causal basis is relatively simple, like the constitutive causal basis of type 1 diabetes and bacterial infections. The survival rate of patients with CML has increased since the introduction of imatinib in 2001, and is now close to the survival rate of people without this form of leukemia. However, although imatinib is effective at targeting CML cells, it is not as magical as classical magic bullets like antibiotics. Although the drug does inhibit an enzyme that exists only in CML cells, it nevertheless has many severe adverse effects (including nausea, diarrhea, headaches, severe cardiac failure, and delayed growth in children). That is because imatinib also inhibits other enzymes that are part of normal physiology, including platelet-derived growth factor, which plays an important role in regulating cell growth and blood vessel formation.[11]

4.4 Non-Specificity

Many medical interventions introduced in recent decades do not satisfy one or both of the magic bullet principles. I have already suggested several reasons why few interventions are magic bullets, and here I more systematically explain this failure. The dearth of magic bullets is a result both of features of interventions and features of

---

[10] The infamous meta-analysis which initially showed this was (Nissen & Wolski, 2007)—I discuss the example of rosiglitazone in Chapter 9. Similarly, although pioglitazone is able to decrease blood sugar levels in patients with type 2 diabetes, major studies have shown it to provide little help in avoiding cardiovascular disease and death (Scheen, 2012).

[11] See (Gambacorti-Passerini et al., 2011). Referring to imatinib, a magazine cover stated: "There is new ammunition in the war against cancer. These are the bullets. Revolutionary new pills like Gleevec combat cancer by targeting only the diseased cells" (*Time*, May 2001). A more careful discussion called it a 'silver bullet' (Pray, 2008). These references to magic bullets overstate the case for imatinib, since it is not true that it targets only cancer cells. Incidentally, because of its patent protection imatinib is also extremely expensive—it costs over $90,000 per year, for the duration of a patient's life.

diseases. In short, many medical interventions do not act specifically, and many diseases have a complex constitutive causal basis that mitigates the effectiveness of interventions—for most drugs, one or both of *Specificity* and *Effectiveness* is not satisfied. Thus, we have few modern magic bullets.

One reason why many medical interventions do not act specifically is that many interventions bind to multiple receptors in various physiological systems (in pharmacodynamic terms, the intervention is not selective). Here is how a pharmacology textbook puts this: "Many clinically important drugs exhibit a broad (low) specificity because the drug is able to interact with multiple receptors in different tissues" (Blumenthal & Garrison, 2011). The textbook gives as an example the drug amiodarone, which is used to treat cardiac arrhythmias. Amiodarone inhibits sodium, calcium, and potassium channels, which modulates the electrical properties of the heart; unfortunately amiodarone also has many harmful effects, perhaps because it is structurally similar to thyroxine (thyroid hormone), which allows it to bind to thyroid receptors.[12] The problem is that a single type of ligand can modulate multiple types of receptors. Moreover, as another pharmacology textbook puts it, "The situation is further complicated by the existence of different receptor subtypes in different tissues" (Dutta, 2009). For example, there are numerous subtypes of serotonin receptors throughout the body, and most ligands with some affinity for a particular serotonin receptor subtype also have some affinity for other serotonin receptor subtypes.

To illustrate how extensive non-selective binding can be, consider the drug aripiprazole (Abilify), a widely prescribed antipsychotic used for treating schizophrenia, bipolar disorder, and depression. Aripiprazole is a ligand for the following receptors: 5-HT_{1A}, 5-HT_{1B}, 5-HT_{1D}, 5-HT_{2A}, 5-HT_{2B}, 5-HT_{2C}, 5-HT_3, 5-HT_{5A}, 5-HT_6, 5-HT_7, D_1, D_2, D_3, D_4, D_5, α_{1A}, α_{1B}, α_{2A}, α_{2B}, α_{2C}, β_1, β_2, H_1, M_1, M_2, M_3, M_4, $M5_1$, among others. This drug is extremely non-selective.

A related problem is that a single type of receptor can be an important component of multiple pathways. Thus, even if a ligand bound selectively to only one receptor, the subsequent physiological effects could be varied. For example, G protein-coupled receptors (GPCRs) are a large class of receptors that bind with ligands and initiate intracellular signaling pathways, which leads to physiological responses. The details of GPCR signaling pathways are incredible. GPCRs are immensely important in human physiology, and are the target of most new pharmaceuticals (Dutta, 2009). When a GPCR is activated, it has a preference for modulating a particular pathway, but most GPCRs can modulate multiple pathways, sometimes as the result of the binding of a single ligand. An example of this is the serotonin receptor 5-HT_{1B}, which is a GPCR involved in multiple pathways throughout the body—5-HT_{1B} is thought to inhibit

---

[12] That is based on what little we know—the textbook notes that "amiodarone's salutary effects and toxicities may also be mediated through interactions with receptors that are poorly characterized or unknown."

the release of dopamine, acetylcholine, and glutamate, all of which are important neurotransmitters involved in a wide variety of neurophysiological activities.

Finally, the same modulated pathway can have very different physiological effects in different tissues. Consider the glutamate pathway, for example. In the liver the glutamate pathway is involved in the metabolism of amino acids, which can lead to urea generation, in the pancreas the glutamate pathway controls insulin secretion by beta cells, and in the brain the glutamate pathway is involved in the modulation of neurotransmitter concentrations.[13] Thus a ligand for glutamate receptors can have wildly different effects on different tissues.

In short, it is common for there to be a one-to-many relationship between ligands and target receptors, a one-to-many relationship between receptors and modulated pathways, and a one-to-many relationship between modulated pathways and physiological effects (at least some of these relationships can also be many-to-many—for example, there are many ligands for the serotonin receptor 5-HT$_{1B}$—but this is tangential to the present point). Given the one-to-many relationships between ligand and receptors, receptor and pathways, and pathway and physiological effects, a single ligand can have a wide range of physiological effects. Drugs have a cascading complexity of effects.[14] This implies that *Specificity* cannot be satisfied by such ligands, and thus such ligands are not magic bullets.

4.5 Complexity

Many diseases have a complex constitutive causal basis. In part for this reason, the way diseases have been characterized has shifted from a monocausal model of disease to a multifactorial model of disease. For many diseases—such as heart disease, type 2 diabetes, and most psychiatric ailments—there are not necessary or sufficient causes but rather a multitude of factors that increase the probability of the disease.[15] If a cause is neither necessary nor sufficient for the disease in question, then intervening on that cause will not be sufficient to eliminate the disease.

A related problem is that many organic systems have a feature called 'robustness.' Robustness (in this context) is the ability of a system to maintain its function when perturbed. Diseases can be robust. Kitano (2007) notes several reasons why robustness can minimize the effectiveness of medical interventions. One is 'systems control':

[13] See (Frigerio, Casimir, Carobbio, & Maechler, 2008). Glutamate receptor ligands are currently being studied as potential treatments for a wide variety of ailments.

[14] Aripiprazole is an example of a drug that is involved at all of these cascading levels: it binds to many receptors, one of which is the 5-HT$_{1B}$ receptor; this receptor activates many pathways throughout the body, including the glutamate pathway; the glutamate pathway has many effects in different tissues. One drug, many effects.

[15] Similarly, there is a complex etiological basis of many diseases. For example, our present understanding of depression involves many kinds of causes, at multiple scales from genes to environmental factors; see (Mitchell, 2009) and (Tabery, 2014). But the complexity of the etiological basis of disease is distinct from the complexity of the constitutive causal basis of disease.

physiological systems, including disease systems, can employ feedback regulation to maintain system function despite perturbation—for example, "the effects of a drug can be neutralized if a negative-feedback loop compensates for changes in the level of the molecule that the drug targets." Another is 'redundancy and diversity' of microphysiological pathways: if one pathway is modulated by a drug, other pathways with similar or overlapping functions can compensate. An example of this is the evolution of bacteria to develop resistance against antibiotics such as penicillin. Robustness is ubiquitous (Walsh, 2015). Thus, it is common for physiological systems, including pathophysiological systems, to have built-in mechanisms that mitigate the effectiveness of medical interventions.

Complicating matters is that for many diseases, notably psychiatric diseases and especially mood disorders, we do not have a good understanding of the constitutive causal basis of the disease (Murphy, 2006). There is a weak epistemic version of this problem and a stronger ontological version. Many such ailments are not in fact diseases, but rather are syndromes (clusters of symptoms). For most such syndromes, it is very likely that the set of people categorized with a syndrome are heterogeneous with respect to the constitutive causal basis of their syndromes (one reason to suspect this is that the categorized people are very heterogeneous with respect to their symptoms).

Take depression, for example. The class of people diagnosed with depression is diverse regarding the symptom profiles of those people. As physicalists we should assume that there is a physical basis of the symptoms that patients diagnosed with depression suffer from. It is likely that the set of people diagnosed with the syndrome depression is in fact composed of people with multiple subtypes of depression—we simply do not know the constitutive causal basis of these subtypes, and thus do not know how to properly partition the set of people diagnosed with depression into subtypes. The monoamine theory of depression, popular in the twentieth century, has fallen out of favor among researchers for a number of reasons: most people diagnosed with depression do not have low serotonin levels, and drugs that modulate serotonin levels are barely effective at modulating depressive symptoms.[16] This is the epistemic problem—we do not know the constitutive causal basis of many diseases. The harder ontological problem is that the symptoms of mood disorders such as depression are mental phenomena, what philosophers call 'intentional' states and 'affective' states. To understand the constitutive causal basis of these symptoms, we would have to understand the constitutive causal basis of these mental states, and arguably of consciousness itself.

This entails that for syndromic categories (such as depression) interventions cannot be magic bullets. That is because *Effectiveness* requires that an intervention target

---

[16] See (Kirsch et al., 2008). Horwitz and Wakefield note that "The major source of evidence for the chemical deficiency hypothesis stems from the success of drug treatments, which raise levels of amines, in alleviating depressive symptoms," but even the originator of the theory, Joseph Schildkraut, recognized that "even if the drugs are effective in treating the disorders, that does not necessarily imply that their mode of action involves correction of the underlying abnormality" (cited in (Horwitz and Wakefield, 2007)). Tsou (2012) gives a more optimistic account of inferring the causal basis of a disease from an intervention's effects.

CAUSAL BASIS OF DISEASE or NORMATIVE BASIS OF DISEASE (or both), but if a set of people is composed of people with multiple diseases (as is the case with syndromic categories), then the set is heterogeneous with respect to the bases of disease, and so an intervention that is specific can at best improve the health of a subset of people (namely, the subset of people who in fact have the causal basis or normative basis that the intervention targets).

One could respond by noting that this is merely an epistemic limitation regarding the proper way to demarcate disease categories; as we develop knowledge of the constitutive causal basis of subtypes of syndromic categories, these categories will come to satisfy the conditions required for a state to be a unique disease, and there is no in-principle barrier to discovering interventions that target such a disease. I opened this chapter with a case that illustrates such a development from psychiatry, namely the discovery of the constitutive causal basis of a subtype of psychosis based on infection with syphilis. Prior to the discovery of neurosyphilis, psychosis was a broad category, which included all those troubled people who had lost contact with reality. Once it was discovered that for some psychotic patients the cause of their psychosis was syphilis infection, there was an identifiable subtype of psychosis for which the causal basis of the disease was known and targetable. Ehrlich's development of arsphenamine provided a magic bullet, which specifically targeted this constitutive causal basis. Thus, goes this response, coarse-grained syndromic disease categories can be refined as microphysiological knowledge of disease states improves, and so there is no principled reason to think that syndromic categories cannot have magic bullets. That the discovery of the constitutive causal basis of neurosyphilis is the only example of this occurring in the history of psychiatry ought to make one cautious about this line of reasoning. In any case, this response does not help address the complexity of the constitutive causal basis of many diseases, and at least for mood disorders does not address the harder ontological problem noted above.

In sum: both the nature of interventions and the nature of diseases implies that many interventions cannot satisfy *Effectiveness* or *Specificity* (or usually both), and thus there are few magic bullets.[17]

4.6 Conclusion

The magic bullet model of medical interventions is comprised of two principles: *Effectiveness* and *Specificity*. This model has been a guiding ideal for medical interventions. Unfortunately, medicine has few magic bullets. This is a result both of the

---

[17] Another factor that could partly explain the dearth of recent magic bullets is industry's focus on 'me-too' drugs. A me-too drug is a new member in a class of drugs for which there already exist several other members. Manufacturers are motivated to develop me-too drugs because they already know that other members in the class of drugs are profitable, and because a me-too drug is similar to others in the class, it will likely get approved by regulators, thus minimizing the risk of investment. Examples of me-too drugs include selective serotonin reuptake inhibitors and statins.

cascading complexity of physiological effects of interventions and of the complexity and robustness of diseases.

In the arguments in Part II my focus is on problems with how medical interventions are tested. These problems are exacerbated by biases in medical research, which themselves are exacerbated by conflicts of interest that arise in the context of corporate research on financially profitable pharmaceuticals. These methodological and social problems are related to the magic bullet model. The kinds of methods that have been devised to evaluate the effectiveness of interventions, most prominently randomized trials, are suited to products that are based on the magic bullet model, namely pharmaceuticals (it is difficult to perform a randomized trial with subject allocation concealment on many surgical interventions, or lifestyle interventions, or social policy interventions). Magic bullets create powerful financial incentives, because they are cheap to mass produce and easy to distribute and sell, as opposed to very different models of interventions for improving health, especially those that in Chapter 12 I call gentle medicine, such as lifestyle interventions or socioeconomic interventions. The latter sorts of interventions, which are not based on the magic bullet model, are not patentable and easy to produce, distribute, and sell. The methodological problems in Part II of this book (malleable methods, which overestimate effectiveness) and social problems in Part III of this book (biased research, which is not driven by real social needs) are in part a result of our excessive focus on developing interventions that attempt to follow the model of the magic bullet.

I am not suggesting that we entirely dispel with the magic bullet model—on the contrary, as I emphasize in §4.1, some of our best medical interventions are good approximations of the magic bullet ideal. The magic bullet model remains an ideal worth striving towards for at least some domains. For example, as our antibiotics become less effective due to the development of antibiotic-resistant organisms, it will be important to develop new magic bullet antibiotics. Nevertheless, we should understand how rare magic bullets are. Given that the majority of our medical interventions fall short of the magic bullet ideal, we should devote more research towards understanding different kinds of interventions for improving health (see Chapter 12).

I began this chapter with a story about Ehrlich's hunt for magic bullets. Ehrlich was explicitly referring to a German folktale called 'Der Freischütz', or 'The Marksman.'[18] In this tale a young man, Max, must win a shooting contest so he can marry Agathe, the woman he loves. But Max is not a good marksman. So he makes a deal with the devil: Max sells his soul to the devil in return for magic bullets, which always hit the desired target. During the contest, when Max fires the magic bullet, the devil guides the bullet into Agathe. Max sold his soul to obtain magic bullets, which only harmed his beloved.

---

[18] The tale was adapted as an opera by Carl Maria von Weber, and premiered in 1821. Drews (2004) makes note of Ehrlich's appeal to this tale in the context of Ehrlich's immunological work (Ehrlich used the term magic bullet to describe antibodies).

PART II
Methods

5

Down with the Hierarchies

5.1 Evidence Hierarchies

One of the great challenges of modern medical research is the assessment of causal hypotheses based on disparate kinds of evidence. For typical hypotheses regarding the effectiveness of pharmaceuticals there is evidence from computational models of toxicity, studies on cells and tissues, experiments on multiple animal species investigating multiple organ systems (murine and canine, and primate and porcine, alive at first, then dead, dissected, and analyzed), multiple kinds of epidemiological studies of human populations, randomized trials, and summaries of the available evidence by formal techniques such as meta-analysis and by social methods such as consensus conferences. Each of these kinds of evidence itself has many variations. Epidemiological studies on humans, for instance, include case-control studies, retrospective cohort studies, and prospective cohort studies.

The evidence-based medicine (EBM) movement has organized and assessed this huge volume and diversity of evidence with evidence hierarchies, which are rank orderings of kinds of research methods according to the potential for those methods to suffer from systematic bias. This ordering is usually determined by one or very few parameters of study designs. Systematic reviews (which often include meta-analyses) are typically held to be at the top of such hierarchies, randomized controlled trials (RCTs) near the top, non-randomized cohort and case-control studies held to be lower, and near the bottom are laboratory studies and anecdotal case reports.

The use of evidence hierarchies is widespread in medical science, and such hierarchies continue to be an organizing epistemological principle for evidence-based medicine.[1] The primary function of evidence hierarchies is to organize and filter the great volume and diversity of evidence when assessing the effectiveness and harms of medical interventions. Standard guidance is to attend only to evidence from the top of evidence hierarchies.

---

[1] The evidence-based medicine movement emphasizes the employment of evidence hierarchies: "wise use of the literature requires a sophisticated hierarchy of evidence" (Karanicolas, Kunz, & Guyatt, 2008). To see some evidence hierarchies described and defended, see (Wilson, Hayward, Tunis, Bass, & Guyatt, 1995), (Hadorn, Baker, Hodges, & Hicks, 1996), and (Atkins, Best, Briss, & Group, 2004).

Table 5.1. Evidence hierarchy for detecting treatment benefits of medical interventions

Level of Evidence	Method
1	Systematic review of randomized trials
2	Randomized trial or observational study with dramatic effect
3	Non-randomized controlled cohort/follow-up study
4	Case series, case-control studies, or historically controlled studies
5	Mechanism-based reasoning

In this chapter I criticize the use of evidence hierarchies. I argue that evidence hierarchies should not be used when assessing medical interventions. Several of the problems with evidence hierarchies have been formulated in detail by others, and I draw on such criticisms in the arguments of §5.2–5.5.[2] Some recent evidence hierarchies have attempted to accommodate some of these previous criticisms, and are thus improvements over the original hierarchies. My argument in §5.6 is that the methodological innovations associated with these recent hierarchies in fact amount to an outright abandonment of evidence hierarchies. Evidence hierarchies are an epistemologically impoverished tool for assessing medical interventions.

Table 5.1 shows a representative example of an evidence hierarchy. This hierarchy, taken from the Oxford Centre for Evidence-Based Medicine, illustrates the categorical rank ordering of types of methods.

A hierarchy is, formally, a partially ordered set, in which the ordering of the elements in the set is based on a particular property.[3] For standard evidence hierarchies the elements that are ordered are different kinds of methods, and the property on which the orderings are based is usually the internal validity of a method relative to hypotheses regarding effectiveness of interventions.[4] Graphical representation of hierarchies, such as Table 5.1, place some elements above or below other elements based on such orderings. There is no absolute significance of the place that a method holds in a hierarchy: not all tokens of methods from the top of an evidence hierarchy are reliable (some randomized trials are seriously flawed, as I argue in §5.5 and in further detail in Chapters 8, 9, and 10), and not all tokens of methods from the bottom of an evidence hierarchy are unreliable. I am not aware of anyone who defends this absolute interpretation of evidence hierarchies. Rather, the usual understanding of evidence hierarchies is a weaker relative interpretation, which holds that tokens of methods from higher in

---

[2] These include (Bluhm, 2005), (Upshur, 2005), (Rawlins, 2008), (Goldenberg, 2009), (Borgerson, 2009), (Solomon, 2011), and (La Caze, 2011). For a specific critique of placing meta-analysis at the top of such hierarchies, see Chapter 6.

[3] Etymologically, a hierarchy refers to 'rule by priests,' in which the hierarch is the top ruling priest.

[4] Internal validity is freedom from systematic bias in a method. This is usually contrasted with external validity, which is the validity of extrapolating results from a test situation to a target situation. For a classic statement of these terms, see (Cook & Campbell, 1979).

an evidence hierarchy are more reliable than tokens of methods from lower in the hierarchy. The relative interpretation is nevertheless a strong position in that it holds that the ordering of methods is categorical: that is, it holds that types of methods from higher in an evidence hierarchy are necessarily more reliable, *ceteris paribus*, than types of methods from lower in the hierarchy. It is this relative but categorical interpretation of evidence hierarchies against which I direct most of my arguments.

The relative categorical interpretation of evidence hierarchies has often been expressed in methodological guidance for medical researchers and physicians studying the research literature.[5] These statements of methodological guidance are not committed to an absolute interpretation of evidence hierarchies, since they are not claiming that all token studies from the top of an evidence hierarchy are high quality, but they are committed to a relative categorical interpretation of evidence hierarchies, because they are claiming that only randomized trials provide reliable evidence for making inferences about the effectiveness of medical interventions. Recent evidence hierarchies dispel with the categorical interpretation of evidence hierarchies, and in §5.6 I argue that this amounts to dispelling with the epistemological structure of evidence hierarchies altogether.

I begin by granting, for the sake of argument, that for a particular type of hypothesis a principled justification can be provided for a particular hierarchy of evidence types. Even if this is granted, however, evidence users employ real evidence tokens rather than ideal evidence types, and the warrant for a hierarchy of ideal evidence types may not apply to real evidence tokens (§5.2). For different hypothesis types, different evidence hierarchies may be warranted (§5.3). As a tool for assessing evidence, hierarchies are crude devices compared to other available tools, which include quantitative scales and checklists based on more relevant parameters than those included in evidence hierarchies (§5.4). The principled reasons proposed for evidence hierarchies are inadequate, but even if we grant that an evidence hierarchy has a well-defined top—a type of method deemed categorically superior to others lower in the hierarchy—among tokens of the type of method often said to be at the top (typically randomized trials, systematic reviews, or meta-analyses), in practice there is wide variability in quality (§5.5). Methodologists have begun to develop more sophisticated hierarchies in an attempt to accommodate some of these criticisms, but in §5.6 I argue that any vestiges of evidence hierarchies that remain should be abandoned. In short, I argue that evidence hierarchies should not be used to assess the effectiveness or harms of medical interventions. The considerations I draw on suggest that medical research, even that which is usually considered to be at the top of evidence hierarchies, is highly malleable (§5.7), a point I develop in more detail in later chapters.

---

[5] For instance, an early EBM article stated that one should "discard at once all articles on therapy that are not about randomized trials" (Department of Clinical Epidemiology and Biostatistics, 1981). Similarly, a textbook claims that "if a study wasn't randomised, we suggest that you stop reading it and go on to the next article in your search" (Straus, Richardson, Glasziou, & Haynes, 2005).

5.2 Evidence Users Employ Evidence Tokens

Even if principled arguments could warrant an evidence hierarchy, that hierarchy would be constituted by ideal types of evidence. Evidence employed by users—policy-makers, physicians, patients—is constituted by evidence tokens: real instantiations of the ideal types. Principled arguments that warrant a hierarchy of evidence types would not necessarily warrant a hierarchy of evidence tokens, because real tokens of evidence do not necessarily possess the properties of ideal types of evidence.

The evidence type typically thought to be near the top of evidence hierarchies—evidence from randomized trials—is powerful, in principle. Cartwright (2010) shows that, given a probabilistic theory of causality and some assumptions about the structure of a trial, if the probability that an effect of interest (E) is achieved in the group in which the purported cause (C) was administered—that is, $P(E|C)$—is greater than the probability of E in the control group in which a placebo or other non-C factor was administered for comparison—that is, $P(E|{\sim}C)$—then the evidence from the trial deductively implies that C causes E. An assumption that is necessary for this inference is that the two groups being compared must be identical with respect to all other causes of E not including C.[6] An ideal randomized trial guarantees this. Other types of evidence employed in medical research, such as case-control studies and laboratory research on pathophysiological mechanisms, lack the ability to deductively entail causal conclusions about effectiveness. This provides some justification for placing ideal randomized trials at the top of evidence hierarchies.

Real randomized trials, on the other hand, do not satisfy the strong assumptions required to show that evidence from a trial deductively entails a causal conclusion. Real trials are flawed in various ways, and these flaws render the assumptions regarding ideal trials unwarranted. One flaw in particular is that randomization does not guarantee that all other causes of E not including C are distributed equally between the experimental group and the control group (Worrall, 2002). Trials can be flawed in many other ways, including a biased operationalization of the outcome of interest, an unrepresentative selection of subjects, and a misleading analysis and presentation of results (see Chapters 8 through 10). In §5.5 I survey some of the arguments showing that real randomized trials do not satisfy the strong assumptions of ideal randomized trials. If the strong ideal assumptions are not warranted, then the evidence from a trial does not deductively entail a causal conclusion. Like other empirical methods, then, a real trial merely provides some fallible inductive evidence that C is a cause of E. In short, arguments that warrant a hierarchy of ideal evidence types do not warrant a hierarchy of real evidence tokens.

---

[6] These other causes are sometimes called confounding causes. A standard probabilistic definition of causality is: C causes E iff $P(E|C \ \& \ K_i) > P(E|{\sim}C \ \& \ K_i)$, where K_i are the confounding causes. See (Cartwright, 1979) for a canonical, but slightly different, definition. For a similar defense of the importance of randomization, see (Papineau, 1994). Fuller (forthcoming) offers an alternative formulation of the requirement that confounds be identically distributed.

For an example of a misleading trial, consider the RECORD trial, which was intended to provide evidence that rosiglitazone (Avandia), a drug for type 2 diabetes, is safe. In the early 2000s some smaller trials had suggested that rosiglitazone causes cardiovascular disease and death. This motivated the manufacturer of rosiglitazone to initiate the RECORD trial. This trial employed numerous criteria to specify the kinds of subjects that could be included in the trial. The result of these criteria was that 99 percent of the subjects were Caucasian (despite the fact that the trial took place in dozens of countries), and the subjects in the trial were healthier than the equivalent demographic in the broader population. After the data from this trial was released, an FDA employee reviewed the trial and argued that it was extremely biased. For example, the trial was not blinded (see Chapter 10 for a discussion of biases in medical research). I return to the example of rosiglitazone and this trial in Chapters 8 and 9.

Real randomized trials have an important methodological safeguard compared with other methods. Selection bias is the intentional or unintentional allocation of subjects into groups of a trial in a way that introduces relevant differences between the groups (confounds) which bias the results of the trial. Random allocation of subjects into groups is a way to avoid selection bias (see Chapter 10). This is an important feature of a method, but does not justify placing more confidence in all tokens of this type of method over tokens of other method types.

5.3 Different Hypothesis Types, Different Hierarchies

Even if principled arguments could justify an evidence hierarchy for a particular kind of hypothesis, these arguments would not necessarily warrant an evidence hierarchy for other kinds of hypotheses. Recall my discussion of scope requirements for effectiveness in Chapter 3 (§3.3). The typical kind of hypothesis for which evidence hierarchies are said to apply is a WORKS SOMEWHERE hypothesis (that is, hypotheses about whether or not an intervention tested in a controlled experimental setting was efficacious). But as I argued in Chapter 3, a more important kind of hypothesis for decision-makers is a WORKS GENERALLY or WORKS FOR ME hypothesis (that is, hypotheses regarding effectiveness of an intervention for a target population or a particular patient who is contemplating treatment). Just as important are hypotheses of the kind CAUSES HARM—that is, hypotheses regarding the harm profile of medical interventions—which usually require types of evidence placed low on evidence hierarchies.

Consider the kind of hypothesis WORKS FOR ME. For such hypotheses it is not enough to have evidence for the different kind of hypothesis WORKS SOMEWHERE. This is a point that Cartwright has argued (2007). The latter kind of hypothesis might be well supported by evidence from a randomized trial. But when implementing the intervention outside the experimental setting, the causal structure of the target population might be significantly different than the causal structure of the experimental situation (this is the problem of external validity). Different kinds of evidence may be required to determine and address this. The evidence hierarchy that may be optimal for *WORKS SOMEWHERE* hypotheses may not apply to *WORKS FOR ME* hypotheses.

We ought to be especially concerned with the kind of hypothesis *CAUSES HARM*. I argue in Chapter 9 that clinical research is especially bad at gathering reliable evidence for such hypotheses. Even if principled arguments could justify an evidence hierarchy with randomized trials or meta-analyses at the top for effectiveness hypotheses, the best evidence for many hypotheses about harms does not come from trials or meta-analyses. This is for numerous reasons. There is a direct trade-off between the ability of trials to detect benefits of medical interventions (providing evidence for a *WORKS SOMEWHERE* kind of hypothesis) and the ability of trials to detect harms of medical interventions (providing evidence for a *CAUSES HARM* kind of hypothesis). This trade-off is generated by multiple properties of trials, including the kinds of subjects included or excluded from a trial, the kinds of parameters measured, and the kinds of analyses performed.[7] The vast majority of randomized trials in medical research maximize the ability to detect benefit at the expense of the ability to detect harm. The majority of serious harms caused by medical interventions are detected by post-approval observational analyses (§9.5). Thus, for *CAUSES HARM* hypotheses, medical research relies on evidence from non-randomized studies.

5.4 Assessing and Amalgamating Evidence

Evidence hierarchies constrain quality comparisons between types of evidence to ordinal rankings based on few parameters. More sophisticated evidence assessment tools—I call these 'quality assessment tools' (QATs)—permit assessments of evidence using numerical scales or checklists, which are based on a large number of relevant parameters of medical studies. In Chapter 7 I describe such evidence assessment tools in more detail.

Many of these detailed evidence assessment tools are now available. Some are simple, evaluating only a handful of parameters of medical studies, while others are more complex, evaluating dozens of parameters of studies. These parameters are about the design of a study (whether or not it is randomized, placebo-controlled, etc.), whether or not the design of the study was relevant to the research question at hand, whether or not allocation of subjects was concealed from subjects and investigators, and whether or not the statistical analyses were appropriate. From these parameters an overall numerical score is computed, which is an estimate of the quality of evidence from a study.

Note the relative sophistication of most QATs compared to the simplicity of evidence hierarchies: these tools are able to account for multiple properties of medical studies, and their structure permits differential weighting of the relative importance

---

[7] See (Vandenbroucke, 2008) and (Osimani, 2014) for a discussion of some of these trade-offs, and for a defense of the view that different kinds of hypotheses might require different evidence hierarchies. See also (Borgerson, 2008).

of properties of medical studies. Moreover, the evaluation of evidence from medical studies using such tools is usually quantitative. In §5.6 I argue that, based on such considerations, even the best evidence hierarchies are not as good as typical QATs. The relative sophistication of these evidence assessment tools should not lead us to overestimate their rigor—in Chapter 7 I describe these tools in more detail and discuss several worrying philosophical implications and practical problems associated with them. Nevertheless, these tools show just how crude evidence hierarchies are, since evidence hierarchies are limited by their structure to assessing medical studies based on a small number of properties, and the assessment itself is limited to an ordinal ranking of kinds of studies.

Evidence hierarchies are supposed to provide epistemic guidance when faced with diverse evidence. However, the guidance associated with evidence hierarchies is typically to ignore evidence generated from methods low on a hierarchy, rather than consider it together with evidence generated from methods higher on the hierarchy. Even if the guidance was to somehow consider the evidence generated from methods low on the hierarchy jointly with evidence generated from methods higher on the hierarchy, the hierarchies themselves lack a substantive technique for doing so. The little formal structure of evidence hierarchies is too crude to adequately compare or amalgamate evidence of diverse kinds. As discussed above, most evidence hierarchies categorically rank method types, with randomized trials or meta-analysis at the top and cohort and case-control studies lower. More recent evidence hierarchies allow the assessment of method quality to be modified based on additional properties of studies. Nevertheless, even these newer hierarchies do not propose a technique for how the evidence from different kinds of studies should be integrated into an overall assessment of effectiveness.[8]

The way that evidence hierarchies are usually applied is by simply ignoring evidence from methods lower on the hierarchies and considering only evidence from RCTs (or meta-analyses). However, evidence from multiple levels of evidence hierarchies should be considered when assessing the effectiveness and harms of medical interventions.[9] When assessing benefits and harms of medical interventions we ought to take into account multiple kinds of evidence, but evidence hierarchies offer no way to manage this.

5.5 Trouble at the Top

At the top of evidence hierarchies in medicine are randomized controlled trials or systematic reviews (which typically involve meta-analyses, discussed in Chapter 6).

---

[8] See (Douglas, 2012) for a discussion of methods for amalgamating diverse kinds of evidence. For an interesting discussion of judgments required in assessing evidence in medicine, see (Kelly & Moore, 2011).

[9] See (Russo & Williamson, 2007), (Illari, 2011), and (Leuridan & Weber, 2011) for arguments emphasizing the importance of mechanisms in warranting causal hypotheses, and for a critical response see (Howick, 2011a).

However, quality among trials and meta-analyses is highly variable. This renders their categorical position at the top of evidence hierarchies dubious.

The best justification for placing randomized trials at the top of evidence hierarchies is that they minimize certain forms of bias, especially selection bias. As discussed above, a key assumption to deductively entail a causal conclusion from a trial is that the two groups being compared must be identical with respect to all other causes of E not including C. If a medical study is prone to selection bias then this assumption is likely not satisfied. Proponents of evidence hierarchies claim that random allocation of subjects to experimental groups guarantees that this assumption is satisfied. However, Worrall (2002) convincingly argues that random allocation of subjects to experimental groups cannot guarantee this. Randomization does not even render this assumption very probable. Rather, a single iteration of randomization makes it likely that some confounding causes (the total composition of which is typically unknown) are more or less balanced between the groups, but not all the confounding causes. Thus, one of the most frequently cited justifications for placing randomized trials near the top of evidence hierarchies is inadequate. Moreover, selection bias is just one of many kinds of bias pervasive in medical research, and so even if a clinical trial could minimize selection bias, it may still have many other systematic biases—I describe some of these biases in Chapters 8, 9, and 10.

Systematic reviews and meta-analyses—specifically meta-analyses of randomized trials—are usually held to be at the top of evidence hierarchies. The intuitive rationale for this is that although one particular trial might be flawed in various ways or might be too small to detect a significant effect of an intervention, when pooled together the flaws in multiple trials might wash out or the pooled sample size might be large enough to detect an effect of the intervention. However, as I argue in detail in Chapter 6, meta-analysis suffers from multiple methodological problems, which render its results liable to be influenced by idiosyncratic and subjective factors. This potential for bias arises because meta-analysis is malleable—multiple contradictory conclusions can be reached on the same hypothesis by different scientists performing their own respective meta-analyses, because there are many aspects of a meta-analysis that are unconstrained and permit wide latitude among analysts.

Thus, the methods normally thought to be at the top of evidence hierarchies in medicine are not as good at minimizing bias and constraining assessments of the effectiveness and harms of medical interventions as they are often made out to be. Randomized trials, meta-analyses, and systematic reviews are malleable.

The mixed quality of trials and meta-analyses has been empirically demonstrated using techniques that are more sophisticated than evidence hierarchies for assessing evidence (namely, the QATs discussed above). These second-order studies have shown a wide disparity in the quality of trials. For example, a research group conducted a systematic review of 107 randomized trials about a particular medical intervention, using three popular QATs. This group found that allocation concealment was unclear in 85 percent of these trials, and that the vast majority of the trials very likely suffered from one bias or another. Another group randomly selected eleven meta-analyses involving 127 trials on medical interventions in various health domains. This group assessed the quality of these trials using QATs, and found the overall quality to be low: for example, only 15 percent of trials reported the method of randomization, and even fewer showed that subject allocation was concealed. Such examples could be easily multiplied.[10]

The notion of a rigid evidence hierarchy is dubious when token examples from the top of evidence hierarchies are so often of low quality. At the very least such examples demonstrate that the absolute interpretation of evidence hierarchies is unjustified. Do such examples show that the relative interpretation of evidence hierarchies is also unjustified? This stronger conclusion does not immediately follow. Recall that the relative interpretation holds only that tokens of methods from higher on an evidence hierarchy are more reliable than tokens of methods from lower on the hierarchy. The mere fact that some randomized trials and meta-analyses are of low quality does not in itself entail that the relative interpretation is unjustified, since it is at least possible that all tokens of those methods typically thought to be lower in evidence hierarchies (e.g., case-control studies) are even less reliable than the above examples of low quality trials. This, however, is extremely implausible. A study could lack the desirable property of randomization but have many other desirable properties, such as large size, proper controls, and subject allocation concealment, and be more compelling than a mediocre randomized trial. There are many tokens of excellent non-randomized studies. For example, the first evidence that suggested that smoking causes lung cancer came from large non-randomized observational studies. In any case, recent evidence hierarchies that try to accommodate more nuanced properties of studies, including the lower quality of some randomized trials, in fact entail a rejection of the structure of hierarchies.

5.6 Abandoning Hierarchies

Some groups have begun to develop evidence hierarchies that appear to dispel with the categorical interpretation. One evidence hierarchy, developed by GRADE (Grading of Recommendations Assessment, Development and Evaluation Working Group), assigns evidence to one of four levels (high, moderate, low, and very low), but these assignments can be changed based on several conditions. Randomized trials are automatically categorized as high quality and observational studies are automatically categorized as low quality, but the level of a trial can be decreased by at least one level (to moderate) if it has methodological problems, and the level of an observational study

---

[10] The citations here are: (Hartling et al., 2011) and (Moher et al., 1998). Some epidemiologists have gone so far as to claim that most published research findings are simply false (e.g., Ioannidis, 2005b, 2008b, 2011), and this includes results of studies from the top of evidence hierarchies.

can be increased by up to two levels (to high) if its evidence is especially striking.[11] It follows that, in principle, this hierarchy permits a ranking that places token observational studies above token trials. Many evidence hierarchies in use today maintain a straightforward commitment to the categorical interpretation, despite the development of more sophisticated hierarchies such as the one described here. At least one evidence hierarchy, though, appears to be non-categorical.

However, even this hierarchy is fundamentally based on the categorical interpretation: its starting point for assessing methods is simply the level assigned to a type of method in the hierarchy, and the possibilities for modifying this assessment based on relevant methodological properties is very limited. The coarse-grained assignment of high quality to a method based on the single property of randomization neglects a vast amount of information that pertains to how reliable that method is, regardless of subsequent modifications to the level assignment (I explain this further below). Moreover, as I also argue below, once one abandons the categorical interpretation of evidence hierarchies, there is reason to think that one has abandoned the structure of hierarchies altogether. The considerations that purport to warrant more sophisticated evidence hierarchies in fact warrant abandoning their use.

Howick (2011b) claims that employing evidence hierarchies is a good strategy in most cases, but he rightly notes that for some medical interventions we have strong confidence in their effectiveness despite a lack of evidence from methods near the top of evidence hierarchies. Howick claims that one should consider "all sufficiently high-quality evidence to be weighted together in support of a hypothesis that a treatment caused a patient-relevant outcome." This is right. However, Howick's suggestion is more revisionary and problematic than he suggests.[12] What constitutes high-quality evidence, and whether or not mechanistic evidence and evidence from non-randomized studies ought to be considered, is precisely what is at issue in debates about evidence hierarchies.

One could interpret Howick's suggestion, and any evidence hierarchy consistent with his suggestion (such as that of GRADE), as an outright abandonment of evidence hierarchies. Howick gives conditions for when mechanistic evidence and evidence from non-randomized studies should be considered, and also suggests that sometimes evidence from randomized trials should be doubted (which, given my arguments in Chapters 8, 9, and 10, is an understatement). If one takes into account methodological nuances of

---

[11] The evidence from an observational study might be especially salient, for instance, if there were a very strong association between the purported cause and its effect, or if there was a distinctive dose-response relationship between purported cause and effect, and there were no obvious threats to the internal validity of the study (see §6.5).

[12] How the various kinds of evidence should be "weighted together" is unstated, and why one ought to consider evidence "in support of a hypothesis" without requiring the consideration of evidence against is presumably an oversight. Howick appeals to the 'principle of total evidence,' but in fact this principle requires one to consider not just high-quality evidence, but all evidence, or at least much more than is typically admitted by evidence hierarchies. In Chapter 6 I also appeal to the principle of total evidence to demonstrate a problem with meta-analysis. See also (Bluhm, 2011).

medical research, then the metaphor of a hierarchy of evidence and its utility in assessing the effectiveness and harmfulness of medical interventions is at best irrelevant.

As I noted above, some evidence hierarchies employ more than one property to rank methods. Trivially, if one develops a ranking of the reliability of methods based on any number of properties, one develops an ordered set, and thus a hierarchy. Thus, at first glance, the ranking of methods based on multiple properties still involves the structure of hierarchies. However, standard evidence hierarchies involve a ranking of *types* of methods (§5.2)—this is what makes them categorical. The employment of multiple properties to rank methods in the way that Howick and the GRADE group suggest involves ranking *tokens* of methods. This is what makes such an approach non-categorical, and thus superior to standard evidence hierarchies. It is not method types (such as randomized trials and case-control studies) that are ultimately being ranked according to these proposals, but rather it is method tokens (this or that particular study) that are being ranked. A system for ranking method tokens is welcome, because, as noted in §5.2, evidence users employ real evidence tokens, not ideal evidence types.

This raises the question of how best to evaluate the quality of method tokens. The use of n number of properties to rank method tokens is formally equivalent to a scoring system based on n properties that discards all information that exceeds what is required to generate a ranking. Scoring systems, such as the QATs discussed in §5.4 and Chapter 7, generate scores that are measured on scales more informative than ordinal scales (such as interval, ratio, or absolute scales). From any measure on a supra-ordinal scale, a ranking can be inferred on an ordinal scale, but not vice versa—from a ranking on an ordinal scale it is impossible to infer measures on supra-ordinal scales (see Appendix 2). The inference from a supra-ordinal measure to an ordinal measure involves discarding any information beyond what is required to generate mere orderings. Thus, the non-categorical method-token hierarchies such as GRADE provide evaluations of evidence that are *necessarily less informative* than evaluations provided by QATs.

There is another important difference between non-categorical method-token hierarchies and most QATs. With non-categorical hierarchies, the evaluation of token methods (that is, particular trials) *begins* with an assignment of quality based on a single property, and that assignment can then subsequently be modified by other properties of that trial. With QATs, on the other hand, token methods are evaluated by the multiple properties deemed relevant and such evaluations are not constrained by the sequential order in which an analyst appeals to a particular property. Moreover, there is in practice a significant difference between the two approaches regarding the number of properties of method tokens that can be evaluated. In principle, the number of properties employed is no different between the two kinds of approaches, though in fact non-categorical method-token hierarchies use a very small number of relevant properties, whereas some QATs employ a large number of relevant properties, which provides for a more detailed evaluation of the evidence.

Another important difference between the two approaches is based on the weight that can be assigned to individual properties of token methods when assessing the overall quality of a method (where weight is the importance of the property relative to other properties with respect to minimizing the chance of inductive error). Because non-categorical hierarchies begin with a quality assignment based on one property and then other properties can only influence the quality assignment by shifting the assignment by a certain number of levels, the weights that can be assigned to various properties relative to other properties are highly constrained. The GRADE hierarchy has four quality levels, and the extent to which a method's level can be switched by a property is limited to one or two levels per property. Thus, the weight assigned to any property p_1 relative to the weight of some other property p_2 is limited to three possible values: the same weight of p_2, half the weight of p_2, or twice the weight of p_2 (depending on whether or not p_1 and p_2 both switched the method's level by one, or if p_1 switched the level by one while p_2 switched the level by two, or vice versa). Thus the weight assigned to each property is highly constrained. With QATs, on the other hand, the relative weight assigned to each property is completely open, and can be set based on rational arguments regarding the respective importance of the various properties, without arbitrary constraints imposed by the structure of the scoring system (Chapter 7).

There is another problem with hierarchies like that of GRADE. A key property of a non-randomized study that determines if its level should be raised are the results from the study. If the results are especially striking, then the level of that study can be raised by one or two levels. Notice that this property is not about the quality or reliability of the study, but is simply determined by the evidence from the study. Crucially, this feature of level-switching is systematically biased toward promoting studies that show significant and large results while denigrating studies that do not show significant and large results. This bias, in turn, can be expected to contribute to overestimating the size of causal relations that are under investigation, precisely because those non-randomized studies that show salient results have their levels increased while those that do not show salient results maintain their low level.

In short, at present some are suggesting a departure from standard evidence hierarchies (categorical method-type hierarchies) in favor of more sophisticated hierarchies (non-categorical method-token hierarchies). This may be due in part to some of the extant criticisms of standard hierarchies canvassed in this chapter. Regardless, any vestiges of evidence hierarchies that remain should be excised, since these vestiges unreasonably constrain evidence assessment, are less informative than other tools for assessing evidence, and contribute to systematic overestimation of the strength of causal relations in medical research.

5.7 Discussion

I have raised several arguments against the use of evidence hierarchies. Nevertheless, modern medical research has an enormous volume and diversity of evidence for many hypotheses, the various kinds of evidence have different inductive strengths and weaknesses, and somehow this messy mass of evidence must be analyzed to provide guidance regarding the effectiveness and harms of medical interventions.

One advantage of evidence hierarchies is that they provide a relatively uniform standard for researchers when assessing this large volume and diversity of evidence. At first glance one might think that this uniformity of standard minimizes what in statistics is called 'researcher degrees of freedom'—the extent to which researchers have unconstrained choices when gathering, assessing, and interpreting evidence. Methods that allow research degrees of freedom are malleable—they can be bent in a variety of ways that introduce bias. Since evidence hierarchies seem to be rigid epistemological structures, they are not very malleable. The central argument of this chapter is that the very aspects of evidence hierarchies that make them seem rigid are poorly justified and render them unreliable tools for assessing quality of evidence. One of the unifying threads of Chapters 5 through 10 is that even the very best methods of medical research—those usually placed at the top of evidence hierarchies—are themselves malleable, and just as worrying, in Chapter 7 I argue that there is a fundamental second-order form of malleability present in evidence assessment tools. Since even the top of evidence hierarchies is constituted by these malleable methods, evidence hierarchies are not in fact as rigid as one might have thought. Even the very best methods in medical research are malleable.

6

Malleability of Meta-Analysis

6.1 Meta-Analysis

Medical scientists are faced with a daunting volume and diversity of evidence for many hypotheses regarding the effectiveness and harmfulness of medical interventions. This avalanche of evidence contributed to the formation of groups dedicated to the systematic review of evidence (such as the Cochrane Collaboration), to journals that publish reviews of existing evidence rather than evidence from original research, and to methods of amalgamating evidence, including social methods, such as consensus conferences, and quantitative methods, such as meta-analysis. My focus in this chapter is on meta-analysis, widely thought to be among the most reliable methods in medical research (see Chapter 5). I describe the purported merits of meta-analysis and the aims that analysts set out to achieve with this method, critically assess the details of the method, and argue that meta-analysis does not generally have the merits that many claim for it. Meta-analysis is malleable.

Meta-analysis is a method that combines evidence from individual studies into summary measures of the beneficial and harmful effects of medical interventions. Many claim that meta-analysis is an especially reliable method (§6.2). I articulate two purported methodological principles that many seem to think meta-analysis is especially good at satisfying: CONSTRAINT—the use of meta-analysis should constrain assessments of medical interventions—and OBJECTIVITY—meta-analysis should be performed in a way that limits the influence of subjective biases and idiosyncrasies of researchers.

I show that the use of meta-analysis often fails to achieve CONSTRAINT (§6.3). Meta-analysis fails to constrain assessments of medical interventions because numerous decisions must be made when performing a meta-analysis, which allow wide latitude for subjective idiosyncrasies to influence the results of a meta-analysis. My argument involves a close examination of these details (§6.4). Meta-analysis is performed by selecting which primary studies are to be included in the meta-analysis, calculating the magnitude of the effect attributed to an intervention for each study, assigning a weight to each study, and then calculating a weighted average of the effect sizes. To help describe the methodology of meta-analysis I draw on the published guidance of the Cochrane Collaboration, an institution of evidence-based medicine that commissions a large number of meta-analyses. Finally, I end by discussing an alternative, older, and arguably better strategy for assessing a large volume and diversity of evidence (§6.5), associated with the epidemiologist Sir Bradford Hill (1897–1991).

There is a debate about whether or not randomized trials are the gold standard of evidence to assess medical interventions.[1] However, it is in fact meta-analysis (or systematic reviews that typically include meta-analyses) that is at the top of the most prominent evidence hierarchies in medicine (see Chapter 5). In what follows I criticize this assumed status of meta-analysis. Meta-analyses, like randomized trials, are malleable, and liable to be influenced by numerous biases. This fact, together with a broader concern about biases in medical research (see Chapter 10), provides support to one of the central premises of the master argument for medical nihilism described in Chapter 11.

6.2 Constraint and Objectivity

The first comprehensive meta-analysis was about extra-sensory perception.[2] Meta-analysis later became the platinum standard of evidence in medicine for several reasons. The sheer volume of available evidence meant that most users of evidence (for example, physicians and policy-makers) could not be aware of all relevant evidence. A proposed solution was to produce systematic reviews of the available evidence. By the 1990s, hundreds of meta-analyses were being published every year, and now thousands are published every year.

Meta-analysis became a prominent method in part due to its purported rigor compared with qualitative and unstructured methods of amalgamating evidence. In contrast with qualitative literature reviews and consensus conferences, meta-analyses have a constrained structure and a quantitative output. The importance of using systematic methods of amalgamating evidence became apparent by the 1970s, when scientists began to review a plethora of evidence with what some took to be personal idiosyncrasies.[3] A recent textbook on meta-analysis worries that unstructured reviews "come to opposite conclusions, with one reporting that a treatment is effective while the other reports that it is not"—the solution to this problem, according to the authors, is to use meta-analysis, a more structured method that (goes this suggestion) can

---

[1] See (Worrall, 2002), (Worrall, 2007), (Borgerson, 2008), (Cartwright, 2007), and (Cartwright, 2009).

[2] (Rhine, Pratt, Stuart, Smith, & Greenwood, 1940). This is a nice historical accident, because Hacking (1988) showed that the practice of randomizing subjects into different groups also began in psychical research—thus both our alleged gold standard of evidence and our alleged platinum standard of evidence first arose in research about paranormal psychology.

[3] An early defender of meta-analysis claimed that "A common method for integrating several studies with inconsistent findings is to carp on the design or analysis deficiencies of all but a few studies—those remaining frequently being one's own work or that of one's students or friends" (Glass, 1976). An example is (Pauling, 1986), in which the Nobel Laureate cited dozens of his own studies supporting his hypothesis that vitamin C reduces the risk of catching a cold, and yet he did not cite studies contradicting this hypothesis, though several had been published (Knipschild, 1994).

constrain assessments of medical interventions.[4] Likewise, a statistics textbook emphasizes a worry regarding reviewers' idiosyncrasies—"the conclusions of one reviewer are often partly subjective, perhaps weighing studies that support the author's preferences more heavily than studies with opposing views"—and the authors suggest that meta-analysis can mitigate this concern.[5]

The scientific basis of meta-analysis is simple. Many purported causes in medicine have a small effect, and so when analyzing data from a single trial on an intervention with a small effect, there might be no statistically significant difference between the experimental group and the control group of the trial. But by pooling data from multiple trials the sample size of the analysis increases, thereby rendering estimates of the magnitude of an intervention's effects more precise, and perhaps statistically significant. A key feature of meta-analysis, then, is quantitative precision, which is especially important for detecting small effects (as I argue in Chapters 8 and 11, many medical interventions have tiny effects).

In short, meta-analysis is a method to amalgamate evidence from multiple studies. Relative to other methods of amalgamating evidence, such as informal reviews or consensus conferences, meta-analysis is said to have the virtues of constraining estimates of the effectiveness and harmfulness of medical interventions and doing so in a way that is not influenced by subjective idiosyncrasies of analysts. The purported rigor, transparency, quantitative precision, and freedom from personal bias can be summarized by these two principles:

CONSTRAINT: Meta-analysis should constrain estimates of the effectiveness and harmfulness of medical interventions.

OBJECTIVITY: Meta-analysis should not be sensitive to idiosyncratic or personal biases.

A straightforward way of construing the relation between these two norms is that OBJECTIVITY is in the service of CONSTRAINT: meta-analysis can constrain estimates of the effectiveness and harmfulness of medical interventions only if it is not sensitive to analysts' idiosyncratic or personal biases.[6] Defenders of meta-analysis claim that, compared with other methods of amalgamating a large volume of evidence, meta-analysis best satisfies these principles. This is the basis of the alleged status of meta-analysis at the top of evidence hierarchies.

However, in the following sections I argue that meta-analysis, unfortunately, generally fails to satisfy these principles. The details of the methodology of a meta-analysis require many decisions at multiple stages, which allow wide latitude for an analyst's idiosyncrasies to affect its outcome. Meta-analysis is malleable.

---

[4] (Borenstein, Hedges, Higgins, & Rothstein, 2009).

[5] Since "it is extremely difficult to balance multiple studies by intuition alone without quantitative tools" (Whitlock & Schluter, 2009), the authors claim meta-analysis should be used.

[6] For a recent historical account of objectivity see (Daston & Galison, 2007), and for a recent philosophical account see (Douglas, 2004).

6.3 Failure of Constraint

Medical scientists have recently noted that multiple meta-analyses about the same medical interventions can reach contradictory conclusions. For example, there have been numerous inconsistent studies on the benefits and harms of synthetic dialysis membrane versus cellulose membrane for patients with acute renal failure: one meta-analysis of these studies found greater survival of such patients using the synthetic membrane compared with those using the cellulose membranes, while another meta-analysis reached the opposite conclusion. Here is another example. Two meta-analyses published in the same issue of *BMJ* came to contradictory conclusions regarding whether or not an association exists between the use of selective serotonin reuptake inhibitors (SSRIs, a class of antidepressants) and suicide attempts. In one, there was no association found between the use of these antidepressants and suicide attempts, and only a weak association between antidepressant use and risk of self-harm, while in the other there was a strong association between antidepressant use and suicide attempts.[7] Contradictory conclusions have been reached from meta-analyses on the benefits of acupuncture and homeopathy, mammography for women under fifty, and the use of antibiotics to treat otitis, to name a few other examples.

Differential outcomes between contradictory meta-analyses can be associated with the analysts' professional or financial affiliations. For example, several meta-analyses have investigated a potential causal relation between formaldehyde and leukemia. Two meta-analyses concluded that formaldehyde exposure does not cause leukemia. In contrast, a third found a modest elevation of risk of developing leukemia in professionals who work with formaldehyde, such as pathologists and embalmers. A fourth found an even higher risk.[8] The meta-analyses that concluded that formaldehyde exposure does not cause leukemia were performed by employees of private consulting and industrial companies. In contrast, the authors of the two meta-analyses that found some evidence for a causal relation between formaldehyde exposure and leukemia worked in academic and government institutions.

Barnes and Bero (1998) performed a quantitative second-order assessment of multiple reviews that reached contradictory conclusions regarding the same hypothesis, and found a very strong correlation between the outcomes of the meta-analyses and the analysts' relationships to industry. They analyzed 106 review papers on the health effects of passive smoking ('secondhand smoke'): thirty-nine of these reviews concluded that passive smoking is not harmful to health, and the remaining sixty-seven concluded that there is some adverse health effect from

---

[7] The four meta-analyses cited here are, respectively: (Subramanian, Venkataraman, & Kellum, 2002), (Jaber et al., 2002), (Gunnell, Saperia, & Ashby, 2005), and (Fergusson et al., 2005).

[8] The citations are: (Bachand, Mundt, Mundt, & Montgomery, 2010), (Collins & Lineker, 2004), (Bosetti, McLaughlin, Tarone, Pira, & La Vecchia, 2008), and (Zhang, Steinmaus, Eastmond, Xin, & Smith, 2009). Formaldehyde exists in products that account for more than 5 percent of the U.S. gross national product (Zhang et al. 2009).

passive smoking. Of the variables investigated, the only significant difference between the analyses that showed adverse health effects versus those that did not was the analysts' relationship to the tobacco industry: analysts who had received funding from the tobacco industry were eighty-eight times more likely to conclude that passive smoking has no adverse health effects compared with analysts who had not received tobacco funding.

Here is another example. Antihypertensive drugs have been tested by hundreds of studies, and as of 2007 there had been 124 meta-analyses on such drugs. Meta-analyses of these drugs were five times more likely to reach positive conclusions regarding the drugs if the reviewer had financial ties to a drug company. Or consider the second-order review of meta-analyses of studies on spinal manipulation as a treatment for lower back pain: some meta-analyses have reached positive conclusions regarding the intervention while other meta-analyses have reached negative conclusions, and a factor associated with positive meta-analyses was the presence of a spinal manipulator on the review team.[9]

Such examples could easily be multiplied. About one third of meta-analyses in medicine are published by employees of the company that manufactures the drug that is assessed by the meta-analysis, and these meta-analyses are twenty times less likely to make negative claims about that drug.[10] The above examples illustrate the fact that multiple meta-analyses of the same primary set of evidence can reach contradictory conclusions. The examples suggest that idiosyncratic features of analysts influence the results of meta-analyses. Moreover, the features of meta-analysis that explain its occasional failure to attain CONSTRAINT are shared by all meta-analyses. That is, the conditions under which multiple meta-analyses of the same primary evidence can reach contradictory conclusions are inherent features of all meta-analyses. To show this I turn to a detailed examination of the method.

6.4 Meta-Analysis is Malleable

The failure of CONSTRAINT in the above cases is at least partially a consequence of the failure of OBJECTIVITY: constraint on assessments of medical interventions was not met by the meta-analyses in §6.3 because the meta-analyses were not sufficiently objective. Subjectivity is infused at many levels of a meta-analysis: when designing and performing a meta-analysis, decisions must be made—based on judgment, expertise, and personal preferences—at each step of a meta-analysis, which include the choice of primary evidence, outcome measure, quality assessment tool, and averaging technique. I examine each choice in turn.

---

[9] The two second-order reviews cited in this paragraph are (Yank, Rennie, & Bero, 2007) and (Assendelft, Koes, Knipschild, & Bouter, 1995).

[10] (Ebrahim, Bance, Athale, Malachowski, & Ioannidis, 2016).

6.4.1 Choice of primary evidence

Multiple decisions must be made regarding what primary evidence to include in a meta-analysis. The dominant view in evidence-based medicine is to include only evidence from randomized trials in a meta-analysis.[11] Such a view excludes other common kinds of evidence, including that from cohort studies and case-control studies, as well as other kinds of evidence that are not in the domain of meta-analyses, such as pathophysiological evidence, evidence from animal experiments, mathematical models, and clinical expertise.

When assessing a medical intervention one should use all available evidence. Consider: an effect size of 2.0x from three randomized trials testing a particular medical intervention should have a different impact on one's assessment of the intervention when considered in the background of fifty case-control studies on the same intervention that show an effect size of 2.2x, versus fifty case-control studies that show an effect size of −0.8x. If one's assessment of the intervention were not different in the two scenarios, one would be making an unreliable inference. One's assessment of a hypothesis after observing new evidence should be guided by all of one's previous evidence (this general norm is called the 'principle of total evidence'), and if it is not then one is liable to make an unreliable inference about the probability that the hypothesis is true in light of the new evidence.

Consider the following guidance from the Cochrane collaboration: "review authors should not make any attempt to combine evidence from randomized trials and [non-randomized studies]" (Cochrane Handbook 13.2.1.1). Such a practice could limit the external validity of a meta-analysis, since randomized trials are typically performed with relatively narrow study parameters while other kinds of evidence—including evidence from non-randomized human studies, studies on animals, and experiments designed to elucidate causal mechanisms, which are often performed on tissue and cell cultures—can have diverse experimental parameters and aid in causal inference.[12]

Even if we grant that randomized trials provide the most reliable evidence, that would not mean that evidence from non-randomized studies is negligible. Indeed, some of our best medical interventions were supported by evidence from non-randomized studies (such as insulin for type 1 diabetes, discussed in Chapter 4), and for many medical interventions we only have evidence from non-randomized studies. A joke in such discussions is that there has never been a randomized trial that has tested the effectiveness of parachutes.

The exclusive use of a narrow range of evidence is purportedly justified by the garbage-in-garbage-out argument: if low-quality evidence is included in a meta-analysis, then the output of the meta-analysis will be low quality. Some take this to entail that

---

[11] For instance, when performing a meta-analysis, Egger, Smith, and Phillips (1997) claim that "researchers should consider including only controlled trials with proper randomisation."

[12] I discuss this further in Chapters 8 and 9. See also (Illari, 2011), (Leuridan & Weber, 2011), (Russo & Williamson, 2007), and (Howick, 2011a).

rather than including all available evidence, meta-analyses should only include the best evidence (Slavin, 1995).

There is something correct about the garbage-in-garbage-out argument, but there are several problems with it. First, as above: if we ignore evidence, even if it comes from a lower-quality method, we violate the principle of total evidence. Second, Worrall (2002), Cartwright (2007), and others have argued that there is no gold standard of evidence; it follows that we ought to take into account evidence of multiple kinds when it is available. Third, the possibility of defeating evidence should compel us to consider all available evidence.[13] Fourth, there is no reason why an analyst cannot assess lower-quality evidence appropriately, simply by assigning a lower weight to such evidence when calculating a weighted average. Fifth, and finally, the veiled premise of the garbage-in-garbage-out argument—that only randomized trials are reliable while non-randomized studies are unreliable—is false. All inductive methods are potentially unreliable.

This last point hints at what is right about the garbage-in-garbage-out argument. If the available evidence from randomized trials suffers from shared systematic biases, then a meta-analysis on that evidence will be systematically biased. Entire domains of medical research suffer from the same systematic biases. For example, all randomized trials on the effectiveness of antidepressants use one of very few scales for measuring the severity of depression, and I argue in Chapters 8 and 9 that such scales are systematically biased toward overestimating the benefits and underestimating the harms of antidepressants—thus any meta-analysis in this domain will be biased. Similarly, any domain in which publication bias is rampant will render meta-analyses in that domain systematically biased.[14] Unfortunately, publication bias is rampant in medicine (Chapter 10).

In short, although all evidence is inductively risky, there are good reasons for including as much evidence as possible in a meta-analysis, though one must be wary of systematically biased evidence. Regardless, when performing a meta-analysis one must make a decision regarding the breadth of methodological quality to include, and this decision can be made differently by different analysts—this is one feature that makes meta-analyses malleable.

Besides methodological quality, there are other properties of medical studies that can vary, and when performing a meta-analysis one must determine the heterogeneity of such properties that one is willing to accept. Some limitation of the diversity of evidence that gets included in a meta-analysis is justifiable. The Cochrane group gives the following proviso: "Meta-analysis should only be considered when a group of studies is sufficiently homogeneous in terms of participants, interventions and outcomes"

---

[13] For example, if Tamara, a specialist in ocean geography, tells me that Kiribati is an island in the Atlantic, then I have some evidence that Kiribati is indeed an island in the Atlantic; but if I later get evidence that Tamara is a compulsive liar then I have lost my reason to believe that Kiribati is an island in the Atlantic. Attending to some of my evidence (Tamara's claim) and ignoring other evidence (about Tamara's honesty) would lead me to believe something false.

[14] See (Jukola, 2015) for a criticism of meta-analysis that focuses on these issues.

(Cochrane Handbook 9.5.1). 'Outcomes' here refers to which parameters are measured; for example, if one study tests the effect of a drug on lowering blood pressure, and another study tests the effect of the same drug on the rate of heart attacks, then there is no shared outcome on which to calculate an average. If multiple studies do not measure the same parameters then there is no sense in calculating an average value of those parameters.

However, sometimes analysts assess heterogeneity among study designs by assessing the statistical variability of the data between studies: high statistical variability, according to this approach, suggests substantive heterogeneity in study designs (the questionable assumption seems to be that a single type of causal relation should generate relatively homogeneous data among subjects in different trials). It would be odd to decide to not perform a meta-analysis simply because of the variability of data between studies, because such data could be produced by a single causal relation that in fact generated variable data. In any case, as the Cochrane group rightly states, deciding whether or not a meta-analysis should be performed requires a judgment regarding the substantive or statistical homogeneity of the relevant studies. Analysts can demarcate the boundary between those studies that are deemed homogeneous and those outside the homogeneous set in a relatively unconstrained manner.

A similar consideration applies to assessing homogeneity of participants and interventions. If we are interested in the effect of a given intervention, we must be consistent with what that intervention is—although a narrow range of intervention diversity (say, using a single dose of an experimental drug) will narrow the range of conclusions one can draw about the intervention. Likewise for the use of a narrow range of participants: before we can know if an intervention works in a broad demographic range, it is reasonable to try to determine if it works in a narrow demographic range. Moreover, some interventions only have a specific effect in a narrow range of subject diversity.[15] Thus, there can be good reasons for limiting the diversity of participants, interventions, and kinds of outcomes to be included in a meta-analysis. In any case, such parameters of meta-analyses are decision points that can influence the outcomes of a meta-analysis.

In sum: there are a plurality of relatively unconstrained decisions regarding what evidence to include that analysts must make when performing a meta-analysis. The worry is that such choices can vary between analysts, and such differences can affect the outcome of a meta-analysis.

Another choice that must be made regarding which primary evidence to include in a meta-analysis is the degree of discordance—that is, the degree to which evidence from different primary studies disagree or contradict each other—that the analyst is willing to accept amongst the primary set of evidence. The Cochrane Handbook has a section that discusses strategies for dealing with discordant evidence (9.5.3). An examination of

---

[15] On the other hand, Epstein (2007) argues that our knowledge of the effectiveness and harm profile of many medical interventions is limited because these interventions have been tested on a narrow demographic range of subjects.

these strategies is revealing. One strategy is to 'explore' the discordance: discordance might be due to systematic differences between studies, and so an analysis can be done to determine if differences between studies are related to differences in outcomes. Another strategy is to exclude studies from the meta-analysis: the handbook claims that discordance might be a result of several outlying studies, and if a factor can be found that might explain the discordance, then those outliers can be excluded. The handbook notes, however, that "Since usually at least one characteristic can be found for any study in any meta-analysis which makes it different from the others, this criterion is unreliable because it is all too easy to fulfill." Indeed, a study can be similar or dissimilar to another in an infinite number of ways, and so if one had sufficient data and resources, one could always find a potential difference-maker about a study. Each of these strategies for dealing with discordance can be pursued in a multitude of ways, with varying amounts of time and energy devoted to the particular strategies. The extent of discordance deemed acceptable in a meta-analysis is something that can be freely decided upon. Differing approaches to discordance can have a direct effect on the outcomes of meta-analyses.

Decisions regarding what primary evidence to include in a meta-analysis are constrained by what primary evidence is available. A well-known problem in medical research is publication bias: studies that show positive findings are more likely to be published than studies that have null or negative findings (Chapter 10). An illustrative example is provided by Whittington et al. (2004), who shows that the risk-benefit profile of some SSRIs for the treatment of childhood depression is positive when considering only published studies but negative when both published and unpublished studies are evaluated. Reviewers performing a meta-analysis often have less access to evidence that suggests that an intervention is ineffective or harmful (because it is unpublished) than they do to evidence that suggests the intervention is effective, and this can influence the results of a meta-analysis because publication bias systematically favors medical interventions.[16]

In sum, a number of decisions must be made regarding which studies to include in a meta-analysis, including the acceptable range of methodological quality of studies, the acceptable range of study parameter diversity, whether or not to exclude studies with outlying data, and if publication bias is severe or not. In terms of the principles described in §6.2, the plurality of required decisions regarding which studies to include in a meta-analysis threatens OBJECTIVITY, and thereby CONSTRAINT. Decisions regarding the choice of primary evidence to be included in meta-analysis must be based on judgment, thereby inviting idiosyncrasy and allowing a degree of latitude in the results of a meta-analysis. This renders meta-analysis malleable.

[16] The issue of which primary studies to include in a meta-analysis is appealed to by analysts when explaining contradictory outcomes between their own meta-analysis and other meta-analyses. For instance, in the report by Bachand et al. (2010)—one of the meta-analyses that tested if formaldehyde exposure causes leukemia, discussed in §6.3—the authors claimed that their finding contradicted that of an earlier meta-analysis because of a difference in selection of primary studies.

6.4.2 Choice of outcome measure

Data from primary studies must be summarized quantitatively by an outcome measure before being amalgamated into a weighted average. An outcome measure is used to compute an effect size, which is an estimate of the magnitude of the purported strength of the causal relation under investigation. Multiple outcome measures can be used for this— including the risk difference, relative risk, and relative risk reduction (I give examples of these below, and discuss outcome measures in detail in Chapter 8, where I criticize the use of relative measures such as the risk ratio, and argue that absolute measures such as risk difference should always be employed). The choice of outcome measure can influence the degree to which the primary evidence appears concordant or discordant, and so ultimately the choice of outcome measure influences the results of meta-analysis.

As discussed above, the Cochrane group gives several strategies for dealing with discordant primary evidence. One of these strategies is to change the outcome measure when faced with discordance. Because of the mathematical relationship between ratios and differences, discordant relative effect sizes can entail concordant absolute effect sizes, and vice versa. A hypothetical case will help me illustrate this.

Consider two studies (1 and 2), each with two experimental groups (E and C), and each with a binary outcome (Y and N). Table 6.1 indicates the possible outcomes for each study, where the letters (a–d) are the numbers for each outcome in each group.

The risk ratio (RR) is defined as:

$$RR = \left[a/(a+b)\right]/\left[c/(c+d)\right]$$

The risk difference (RD) is defined as:

$$RD = a/(a+b) - c/(c+d)$$

Now, suppose for Study 1 the numbers for the two outcomes in each group are a = 1, b = 1, c = 1, d = 3 and for Study 2 they are a = 6, b = 2, c = 3, d = 5. This would give the following effect sizes for the two studies:

RR of study 1 = 2

RR of study 2 = 2

RD of study 1 = 0.25

RD of study 2 = 0.375

Table 6.1. A 2 × 2 table for defining binary outcome measures

	Group	Outcome
	Y	N
E	a	b
C	c	d

Thus, these two studies, using risk difference as the outcome measure, have discordant effect sizes (0.25 and 0.375); but by switching the outcome measure to risk ratios the studies have concordant effect sizes (2 and 2). Although the Cochrane group advises changing the outcome measure if the primary studies have discordant effect sizes, choosing between outcome measures on the basis of trying to avoid discordance is ad hoc. Although it may be true that evidence from multiple studies appears discordant only because one outcome measure is used rather than another, it might not be true: discordance might simply be due to a lack of systematic effect by the intervention.

More to the point, the choice of outcome measure is another decision in which personal judgment is required, and the fact that there are multiple outcome measures allows a range of possible outputs for any meta-analysis. Again, this threatens OBJECTIVITY, since some analysts might choose to change their outcome measure when the primary evidence appears discordant using the originally chosen outcome measure, while other analysts might resist such switching. One's choice of outcome measure has a direct influence on the outcome of a meta-analysis, and thus differing choices of outcome measures directly threatens CONSTRAINT.

6.4.3 Choice of quality assessment tool

Analysts often attempt to account for differences in the size and methodological quality of studies included in a meta-analysis by weighing the studies with a quality assessment tool (QAT).[17] The conclusion of a meta-analysis depends on how the primary evidence is weighed, because the weights are used as a multiplier when the effect sizes are averaged.

There are many features of evidence that should influence how primary evidence is weighed, including features that are relevant to both the internal validity and the external validity of studies. In Chapter 7 I argue that scientists lack principles to determine how these features should be weighed relative to each other. The trouble is that different weighing schemes can give contradictory results when evidence is amalgamated. An empirical demonstration of this was given by a research group that amalgamated data from seventeen trials testing a particular intervention, using twenty-five different tools to assess study quality (thereby effectively performing twenty-five meta-analyses). These quality assessment tools varied in the number of assessed study attributes, from a low of three attributes to a high of thirty-four, and varied in the weight given to the various study attributes. The results were troubling: the amalgamated effect sizes between these twenty-five meta-analyses differed by up to 117 percent —*using exactly the same primary evidence.*[18]

---

[17] In Chapter 5 I argue that QATs are superior to evidence hierarchies for assessing evidence, but in Chapter 7 I focus on QATs and argue that they face their own fundamental problems.

[18] Reported in (Juni, Witschi, Bloch, & Egger, 1999). The authors concluded that "the type of scale used to assess trial quality can dramatically influence the interpretation of meta-analytic studies." In Chapter 7 I note more demonstrations of low inter-tool reliability of QATs.

Not only does the choice of quality assessment tool dramatically influence the results of meta-analysis, but so does the choice of analyst using these tools. A quality assessment tool known as the 'risk of bias tool' was devised by the Cochrane group. To test this tool, Hartling et al. (2009) distributed 163 manuscripts of randomized trials among five reviewers, who assessed the quality of the trials with this tool. They found the inter-rater agreement of quality assessments to be very low. In other words, even when given a single quality assessment tool, and a narrow range of methodological diversity, there was a wide variability in assessments of trial quality.

In short, when performing a meta-analysis, analysts must choose a quality assessment tool and apply the tool to the assessment of particular primary-level studies. The choice of quality assessment tool and variations in the assessments of quality by different analysts violate OBJECTIVITY, and this threatens CONSTRAINT: differing decisions regarding one's quality assessment tool lead to contradictory outcomes of a meta-analysis.

6.4.4 Choice of averaging technique

Once effect sizes are calculated for each study, two common ways to determine the average effect size are possible: sub-group averages and pooled averages. In a pooled average, all subjects from the included studies are merged in the analysis as if they were part of one large study with no distinct demographic sub-groups. A problem with the pooled average approach is that different demographic groups might respond differently to an intervention. For example, a drug might, on average, have a large benefit to males and a small harm to females, and if data from these groups were combined in a pooled average we would erroneously conclude that the drug has, on average, a small benefit to all people, including females.

Maintaining distinct sub-groups in a meta-analysis, which the Cochrane group rightly advises, is an attempt to avoid such problems. However, to determine a sub-group average, either the sub-groups must be consistently demarcated amongst primary studies, or the patient-level data necessary to demarcate sub-groups, such as age and gender, must be available to the analyst. The former is often not the case and the latter is often not available. However, if patient-level demographic data is available, then the analyst can demarcate sub-groups any way she wishes until she finds something interesting, but of course such retrospective data-dredging is liable to support spurious findings (see Chapter 10 for a discussion of this practice, sometimes called p-hacking). More to the point: the choice of average type—pooled or sub-group, and if the latter, the choice of sub-groups—is another decision point in meta-analysis that threatens OBJECTIVITY and CONSTRAINT. It is another feature that makes meta-analysis malleable.

6.5 The Hill Strategy

An older tradition of evidence in medicine, associated with the epidemiologist Sir Bradford Hill, provides a more compelling way to consider the variety of evidence in medical research. Hill was one of the epidemiologists involved in the first large case-control studies during the 1950s that showed a correlation between smoking and lung cancer.[19] The statistician Ronald Fisher had noted the absence of controlled experimental evidence on the association between smoking and lung cancer. Fisher's infamous criticism was that the smoking-cancer correlation could be explained by a common cause of smoking and cancer: he postulated a genetic predisposition that could be a cause of both smoking and cancer, and so he argued that the correlation between smoking and cancer did not show that smoking caused lung cancer. The only way to show this, according to Fisher, was to perform a controlled experiment; of course, for ethical reasons no such experiment could be performed. Hill responded by appealing to a plurality of kinds of evidence that, he argued, when taken together made a compelling case that the association was truly causal.

The evidence that Hill cited as supporting this causal inference was: strength of association between measured parameters; consistency of results between studies; specificity of causes (a specific cause has a specific effect); temporality (causes precede effects); a dose-response gradient of associations between parameters; a plausible biological mechanism that can explain a correlation; coherence with other relevant knowledge, including evidence from laboratory experiments; evidence from controlled experiments; and analogies with other well-established causal relations.[20] Hill considered these as inferential clues, or as epistemic desiderata for discovering causal relations. Although Hill granted that no single desideratum was necessary or sufficient to demonstrate causality, he claimed that jointly the desiderata could make for a good argument for the presence of a causal relation.[21] The important point for the purpose of contrast with meta-analysis is the plurality of reasons and sources of evidence that Hill appealed to.

The desiderata appealed to by Hill depend on diverse kinds of evidence, which lack a shared quantitative measure, so that the evidence cannot be combined by a simple weighted average of numerical effect sizes. Versions of the problems I raised for meta-analysis apply to Hill's approach—especially the choice of primary evidence to include, the choice of measures to quantify the evidence (at least, the evidence that can be quantified), the choice of a quality assessment scale to assess or weigh the evidence, and the choice of averaging technique—these are troublesome for the Hill strategy. But this strategy can at least be used to consider all available evidence, at least in principle.

One can have evidence that satisfies only some of the desiderata while still having ample justification for causal inference. Moreover, unlike meta-analysis, there is no

---

[19] (Doll & Hill, 1950, 1954).

[20] Meta-analysis can be thought of as a formal technique to assess the 'consistency' criterion. Framing meta-analysis this way shows just how much meta-analysis neglects, but also shows that it can be a useful technique nevertheless.

[21] See (Doll, 2003). Howick, Glasziou, and Aronson (2009) restructure Hill's desiderata, and Rothman and Greenland (2005) offer a brief discussion of each of the desiderata. Woodward (2010) more thoroughly discusses the specificity desideratum. Of these desiderata, temporality is plausibly a necessary condition for a causal relation.

simple algorithm to amalgamate the diverse forms of evidence in Hill's approach. There is, then, malleability in the Hill strategy. As I argued above, meta-analysis itself is malleable. The complexity of assessing and amalgamating a large volume and diversity of evidence might inevitably require malleable methods. But the Hill strategy is more constraining than meta-analysis in some respects. If a meta-analysis supports a hypothesis while most of Hill's desiderata provide evidence against the hypothesis, this ought to warrant serious reservation in this hypothesis. Conversely, if most of the desiderata coherently support a particular hypothesis, this is suggestive that the hypothesis is roughly correct.[22] Endorsing the Hill strategy, then, does not necessarily mean endorsing a more tolerant or relaxed attitude toward amalgamating evidence. However, nothing very general can be said regarding when the satisfaction of the desiderata is sufficient to infer causality—the Hill approach requires judgment, just as meta-analysis requires. Both approaches to amalgamating evidence are malleable.

6.6 Conclusion

I have argued that meta-analyses fail to adequately constrain assessments of the effectiveness of medical interventions. This is because the numerous decisions that must be made when designing and performing a meta-analysis require judgment and expertise, and allow biases and idiosyncrasies of reviewers to influence the outcome of the meta-analysis. The failure of *OBJECTIVITY* at least partly explains the failure of *CONSTRAINT*: the many judgments required for meta-analysis explain how multiple meta-analyses of the same primary evidence can reach contradictory conclusions.

There are better and worse ways to perform a meta-analysis. Though I have used the published guidance from the Cochrane group to frame my criticisms, this group has worked to improve the quality of meta-analyses.[23] I appeal to meta-analyses throughout this book.[24] Meta-analysis, when done well, is a valuable method in medical research.

Nevertheless, the general epistemic importance given to meta-analysis is unjustified, since it is so malleable: meta-analysis allows unconstrained choices to influence its results, which in turn explains why the results of meta-analyses are unconstrained. The upshot, one might claim, is merely to urge the improvement of the quality of meta-analyses in ways similar to that already proposed by evidence-based medicine

[22] For instance, in §6.3 I discussed meta-analyses that tested whether formaldehyde exposure causes leukemia. One of these (Zhang et al., 2009) concluded that formaldehyde exposure is indeed associated with leukemia, and in addition to this analysis the authors proposed possible causal mechanisms meant to undergird the outcome of their meta-analysis, thereby appealing to the coherence and plausibility desiderata.

[23] Meta-analyses that are not performed by Cochrane collaborators are twice as likely to have positive conclusions compared with meta-analyses performed by Cochrane collaborators (Tricco, Tetzlaff, Pham, Brehaut, & Moher, 2009). Assuming that Cochrane meta-analyses were higher quality than non-Cochrane meta-analyses (a generally safe assumption), it follows that better meta-analyses are less likely to have a positive conclusion regarding a medical intervention.

[24] For instance, in Chapter 9 I cite a prominent meta-analysis which shows that the drug rosiglitazone causes serious harms (Nissen & Wolski, 2007), and in Chapter 11 I cite another prominent meta-analysis which shows that SSRIs are nearly ineffective for treating depression (Kirsch et al., 2008).

methodologists, in order to achieve more constraint. However, my discussion of the many particular decisions that must be made when performing a meta-analysis indicates that such improvements can only go so far. For at least some of these decisions, the choice between available options is arbitrary; the various proposals to enhance the transparency of reporting of meta-analyses are unable, in principle, to referee between these arbitrary choices (in Chapter 7 I argue that this is the case for many aspects of medical research generally).

One of the criticisms I raised against meta-analysis is its reliance on a narrow range of evidential diversity. An older tradition of evidence in medicine, associated with Sir Bradford Hill, is in this respect superior. However, there is no structured method for assessing, quantifying, and amalgamating the very disparate kinds of evidence that Hill considered. Thus the Hill strategy lacks the apparent objectivity, methodological simplicity, and quantitative output of meta-analysis. But given the central argument of this chapter, the fact that the Hill strategy lacks a simple method of objectively amalgamating diverse evidence is not a strike against it relative to meta-analysis, since I have argued that the objectivity of the latter is a chimera. Both approaches to amalgamating evidence in medicine are malleable.

7

Assessing Medical Evidence

7.1 Quality Assessment Tools

To determine how compelling the available evidence is for a hypothesis, one must take into account substantive details of the methods that generated that evidence. Methodological quality is the extent to which the design, conduct, analysis, and report of a study minimizes potential bias and error (I describe some of these biases in Chapter 10). Methodological quality is a multi-dimensional property that one cannot simply intuit, and so formalized tools have been developed to aid in the assessment of the quality of evidence in medical research. Evidence in medicine is often assessed rather crudely by rank-ordering types of methods according to evidence hierarchies (see Chapter 5). More fine-grained tools that I call quality assessment tools (QATs) have been introduced to assess the quality of medical studies.

In this chapter I examine the use of these codified tools for assessing evidence in medical research. I begin by describing the general properties of these tools, including the methodological features that many of them share and how they are typically employed.[1] I then discuss empirical studies that test their inter-rater reliability (§7.2) and inter-tool reliability (§7.3): they are not very good at constraining intersubjective assessments of methodological quality, and more worrying, the use of different QATs to assess the same primary evidence leads to widely divergent quality assessments of that evidence. This is an instance of a more general problem I call the underdetermination of evidential significance, which holds that in a rich enough empirical situation, the quality of the evidence is underdetermined (§7.4). Nevertheless, I defend the use of these quality assessment tools in medical research, despite the fact that these tools are malleable.

In other chapters of Part II of this book I emphasize the malleability of the best first-order empirical methods in medical research: randomized trials and meta-analyses. In this chapter I show that these second-order tools, which are designed to assess the quality of first-order methods, are also malleable. Thus, the malleability of empirical research in medicine is deep: in addition to the malleability of first-order empirical methods, there is malleability in the second-order tools used to evaluate the first-order methods.

---

[1] For a survey of the most prominent QATs, see (West et al., 2002), and see (Olivo et al., 2008) for an empirical critique.

A quality assessment tool can be either a scale with elements that receive a quantitative score representing the extent to which each element is satisfied by a medical study, or else it can be simply a checklist with elements that are marked as either present or absent in a medical study. Given the emphasis on randomized trials in medical research, many QATs are designed for the evaluation of randomized trials, although there are quality assessment tools for observational studies and systematic reviews. Most share several elements, including questions about how subjects were assigned to experimental groups in a study, whether or not the subjects and clinical scientists were concealed to the subjects' treatment protocol, whether or not there was a sufficient description of subject withdrawal from the study groups, whether or not particular statistical analyses were performed, and whether or not a report of a study disclosed financial relationships between investigators and companies.

The first QAT to be developed, known as the Chalmers scale, was published in 1981. By the mid-1990s there were over two dozen of these tools available, and by 2002 a review identified sixty-eight of them for randomized trials or observational studies.[2] Some are designed for the evaluation of any medical study, while others are designed to assess studies from a particular medical sub-discipline. Some are designed to assess the quality of a study itself, while others are designed to assess the publication from a study, and many assess both.

Quality assessment tools are used for several purposes. When performing a systematic review of the available evidence for a hypothesis, they help reviewers take the quality of the relevant evidence into account. This is typically done in one of two ways. QAT scores, or quality scores, can be used to generate a weighting factor for meta-analysis. Meta-analysis involves calculating a weighted average of effect sizes from individual medical studies, and the weighting of effect sizes can be determined by the quality score of the respective study.[3] Alternatively, such scores can be used as an inclusion criterion for a systematic review, in which any primary-level study that achieves a score above a minimum threshold is included in the systematic review (and any study that does not is excluded). These scores also can be used for purposes not directly associated with a particular systematic review or meta-analysis, but rather to investigate relationships between quality scores and other properties of medical studies. For instance, several second-order studies suggest that there is an inverse correlation between quality score of a trial and effect size measured in that trial—in other words, higher-quality trials tend to have lower estimates of the effectiveness of medical interventions.[4]

Why should medical scientists use such tools to assess evidence? Suppose your evidence seems to support a hypothesis, H_1. But then you learn that there is a systematic error in the method that generated your evidence. Taking into account this systematic

---

[2] See, respectively, (Chalmers et al., 1981), (Moher et al., 1995), and (West et al., 2002).
[3] See Chapter 6 for a detailed discussion of meta-analysis, and Chapter 8 for a detailed discussion of outcome measures and effect sizes.
[4] See, for example, (Moher et al., 1998), (Balk et al., 2002), (Hempel et al., 2011), and (Kaplan & Irvin, 2015).

error, the evidence no longer supports H_1 (perhaps instead the evidence supports a competitor hypothesis, H_2). Had you not taken into account the fine-grained methodological information regarding the systematic error, you would have unwarranted belief in H_1. So you ought to take into account fine-grained methodological information. QATs are attempts to codify such fine-grained methodological information.

Any theory of scientific inference should be sensitive to the importance of methodological details. For instance, Mayo's notion of 'severe testing,' broadly based on aspects of frequentist statistics, requires taking into account fine-grained methodological details. The Severity Principle, to use Mayo's term, claims that, for evidence e, "passing a test T (with e) counts as a good test of or good evidence for H just to the extent that H fits e and T is a *severe test* of H" (1996). Attending to fine-grained methodological details to ensure that one has minimized the probability of committing an error is central to ensuring that the test in question is severe, and thus that the Severity Principle is satisfied. The employment of tools like QATs to take into account detailed information about the methods used to generate evidence ought to seem not just reasonable but necessary based on any theory of scientific inference.

One of the simplest of these codified tools is the Jadad scale, developed in the 1990s to assess clinical studies in pain research (Jadad et al., 1996). Here it is, in full:

1. Was the study described as randomized?
2. Was the study described as double blind?
3. Was there a description of withdrawals and dropouts?

A 'yes' to question 1 and question 2 is given one point each. A 'yes' to question 3, in addition to a description of the number of withdrawals and dropouts in each of the trial sub-groups, and an explanation for the withdrawals or dropouts, receives one point. An additional point is given if the method of randomization is described in the paper, and the method is deemed appropriate. A final point is awarded if the method of subject allocation concealment ('blinding') is described, and the method is deemed appropriate. Thus, a trial can receive between zero and five points on the Jadad scale. This scale has the virtue of being easy to use—it takes about ten minutes to complete for each study—which is helpful if a reviewer must assess hundreds of studies for a particular hypothesis. On the other hand, it is simplistic, and despite its simplicity and ease of use it has low inter-rater reliability (§7.2).

In contrast, the Chalmers scale has thirty questions in several categories, which include the trial protocol, the statistical analysis, and the presentation of results. Similarly, the QAT developed by Cho and Bero (1994) has twenty-four questions (I will call this the Cho scale). At a coarse grain, some of the features of the latter two tools are similar to the features of the Jadad scale: they both include questions about randomization, subject allocation concealment, and subject withdrawal. In addition, these more detailed tools include questions about statistical analyses, control subjects, and other features relevant to minimizing systematic error. The Cho scale and Chalmers scale usually take around thirty to forty minutes to complete for each study. Despite

Table 7.1. Number of methodological features used in six QATs, and weight assigned to three prominent features (adapted from (Jüni et al., 1999))

Scale	Number of Items	Weight of Randomization	Weight of Blinding	Weight of Withdrawal
Chalmers	30	13.0	26.0	7.0
Jadad	3	40.0	40.0	20.0
Cho	24	14.3	8.2	8.2
Reisch	34	5.9	5.9	2.9
Spitzer	32	3.1	3.1	9.4
Linde	7	28.6	28.6	28.6

their added complexity, their scoring systems are kept as simple as possible. For instance, most of the questions on the Cho scale allow only the following answers: 'yes' (2 points), 'partial' (1 point), 'no' (0 points), and 'not applicable' (0 points).

Although certain core methodological features are assessed by most QATs, the relative weight of the overall score given to these features differs widely between these tools. Table 7.1 lists the relative weight of three core features—subject randomization, subject allocation concealment ('blinding'), and subject withdrawal—for six of these codified tools. I refer to each QAT by the surname of the first author of the corresponding publication.[5]

Note two aspects of Table 7.1. First, the number of methodological features assessed by these QATs is highly variable, from 3 to 34. Second, the weight given to particular features is also highly variable. Randomization, for instance, constitutes 3.1 percent of the overall score on the Spitzer scale, whereas it constitutes 40 percent of the Jadad scale. Blinding constitutes 3.1 percent of the Spitzer scale, 5.9 percent of the Reisch scale, and 28.6 percent of the Linde scale. The differences between QATs explains the low inter-tool reliability, which I describe in §7.3. But first I note their low inter-rater reliability.

7.2 Inter-Rater Reliability

The extent to which multiple users of the same rating system achieve similar ratings is referred to as inter-rater reliability. Empirical studies have shown a wide disparity in scores when a QAT is used to assess the same studies by multiple reviewers. The inter-rater reliability of these quality assessment tools is poor.

Evaluations of the inter-rater reliability of a QAT are simple: give publications of studies to multiple reviewers who have been trained to use the QAT, and compare the quality scores assigned by these reviewers to each other. A statistic called kappa (κ) is often computed which provides a measure of agreement between the quality scores

[5] In addition to the first three (already cited), the corresponding citations are: (Reisch, Tyson, & Mize, 1989), (Spitzer et al., 1990), and (Linde et al., 1997).

from the multiple reviewers (although other statistics measuring agreement can also be used).[6] Sometimes the authors of the manuscript and the journals where the manuscripts were published are concealed from the evaluators, sometimes they are not, and sometimes both concealed and non-concealed evaluations are performed to assess the effect of knowing this information. In some tests of inter-rater reliability of QATs the manuscripts all pertain to the same medical intervention, while in other cases the manuscripts pertain to various subjects within a particular medical sub-discipline.

For example, a group assessed the inter-rater reliability of the Jadad scale, using four reviewers to evaluate the quality of seventy-six manuscripts of randomized trials.[7] Inter-rater reliability was found to be poor, but it increased when the third item of the scale (explanation of withdrawal from study) was removed and only the remaining two questions were employed. In another study a group tested a QAT called the 'risk of bias tool' for its inter-rater reliability. This group distributed 163 manuscripts of randomized trials among five reviewers, who assessed the trials with this tool, and they found the inter-rater reliability of the quality assessments to be very low. In another study a group used three QATs to assess 107 studies on a medical intervention. Two independent reviewers scored the 107 studies using the three QATs. They found that inter-rater reliability was 'moderate.' However, this was based on a standard scale in which a κ measure between 0.41 and 0.6 is deemed moderate. The κ measure in this paper was 0.41, so it was just barely within the range deemed moderate. The next lower category, with a κ measure between 0.21 and 0.4, is deemed 'fair' by this standard scale. But at least in the context of measuring inter-rater reliability of QATs, a κ of 0.4 represents wide disagreement between reviewers.[8]

Here is a toy example to illustrate the disagreement that a κ measure of 0.4 represents. Suppose two teaching assistants, Beth and Sara, are grading the same class of 100 students, and must decide whether or not each student passes or fails. Their distribution of grades is given in Table 7.2.

Of the 100 students, they agree on passing forty students and failing thirty others, thus their frequency of agreement is 0.7. But the probability of random agreement is 0.5, because Beth passes 50 percent of the students and Sara passes 60 percent of the students, so the probability that Beth and Sara would agree on passing a randomly chosen student is 0.5 × 0.6 (= 0.3), and similarly the probability that Beth and Sara would agree on failing a randomly chosen student is 0.5 × 0.4 (= 0.2), and so the overall

---

[6] For simplicity I will describe Cohen's kappa, which measures the agreement of two reviewers who classify items into discrete categories, and is computed as follows:

$$\kappa = \left[p(a) - p(e) \right] / \left[1 - p(e) \right]$$

where p(a) is the observed frequency of agreement and p(e) is the probability of chance agreement (also calculated from observed frequencies). Kappa was introduced by Cohen (1960). I give an example of a calculation of κ below.

[7] I noted several of these examples in Chapter 6.

[8] The three citations in this paragraph are: (Clark et al., 1999), (Hartling et al., 2009), and (Hartling et al., 2011).

Table 7.2. Illustration of kappa measure:
extent of agreement on grades assigned
by two teaching assistants

		Sara	
		Pass	Fail
Beth	Pass	40	10
	Fail	20	30

probability of agreeing on passing or failing a randomly chosen student is $0.3 + 0.2 = 0.5$. Applying the kappa formula gives: $(0.7 - 0.5)/(1 - 0.5) = 0.4$.

Beth and Sara disagree about thirty students regarding a simple property (passing). It is natural to suppose that they disagree most about 'borderline' students, and their disagreement is made stark because Beth and Sara have a blunt evaluative tool (pass/fail rather than, say, letter grades). But a finer-grained evaluative tool would not mitigate such disagreement, since there would be more evaluative categories about which they could disagree for each student; a finer-grained evaluative tool would increase, rather than decrease, the number of borderline cases (because there are borderline cases between each letter grade). A κ score is a relatively arbitrary measure of disagreement, and the significance of the disagreement that a particular κ score represents presumably varies with context.[9] This example is meant to illustrate that a κ measure of 0.4 represents poor agreement between two reviewers. Nevertheless, I hope that this example illustrates the extent of disagreement found in empirical assessments of the inter-rater reliability of QATs.

In short, different users of the same QAT, when assessing the same evidence, generate diverging assessments of the quality of that evidence. In most of these studies the methods being assessed are of a single design (RCTs), and are about a narrow range of subject matter (usually all the trials are about the effectiveness of a particular medical intervention). The poor inter-rater reliability is even more striking considering the narrow range of study designs and subject matter.

7.3 Inter-Tool Reliability

The extent to which multiple instruments have correlated measurements when applied to the same property being measured I will call inter-tool reliability. One quality assessment tool has inter-tool reliability with respect to another if its measurement of the quality of medical studies correlates with measurements of the quality of the same

---

[9] A κ measure can not only seem inappropriately low, but in some cases can seem inappropriately high. If a κ measure approaches 1, this might suggest agreement that is too good to be true. If Beth and Sara had a very high a κ measure, then one might wonder if they colluded in their grading. I thank Jonah Schupbach for noting this. When using a κ measure to assess inter-rater reliability, the range of measures that indicates agreement is context-sensitive.

studies by the other tool. Quality scores are measures on relatively arbitrary scales, and scores between multiple QATs are typically incommensurable, so constructs such as 'high quality' and 'low quality' are developed, which allow the scores from different tools to be compared. When testing the inter-tool reliability of multiple QATs, what is usually being compared is the extent of their agreement regarding the categorization of particular trials into defined bins of quality. Empirical evaluations of the inter-tool reliability of QATs have shown a wide disparity of quality scores when applied to the same primary-level studies; that is, the inter-tool reliability of quality assessment tools is poor. There have not been many such assessments, though, and those published thus far have varied with respect to the particular QATs assessed, the design of the reliability assessment, and the statistical analyses employed.[10]

An extensive investigation of inter-tool reliability of QATs was reported by Jüni et al. (1999). They amalgamated data from seventeen studies which had tested a particular medical intervention, and they used twenty-five quality assessment tools to assess the seventeen studies (in effect this group performed twenty-five different meta-analyses on the same seventeen studies).[11] Their results were troubling: the calculated effect sizes between these twenty-five meta-analyses differed by up to 117 percent, and they found that medical trials deemed high quality according to one quality assessment tool could be deemed low quality according to another.

Another evaluation of inter-tool reliability of QATs was reported by Hartling et al. (2011), discussed in §7.2. Recall that this group used three QATs to assess 107 trials on a particular intervention. In addition to their finding that inter-rater reliability was low, they also found that the inter-tool reliability was very low. Yet another example of a test of inter-tool reliability of quality assessment tools used six QATs to evaluate twelve trials of a particular medical intervention, and again, the inter-tool reliability was found to be low.[12]

Low inter-tool reliability of QATs is troubling: it is a quantitative empirical demonstration that the assessed quality of a trial depends on the choice of quality assessment tool. In §7.1 I noted that there are many such tools available, and between them there are substantial differences in their design. Thus, the *best* tools that medical scientists have to determine the quality of evidence generated by what are typically deemed the *best* study designs (randomized trials) are relatively unconstraining and liable to produce conflicting assessments of methodological quality. Such low inter-tool reliability of QATs has important practical consequences. In Chapter 6 I showed that multiple meta-analyses of the same primary evidence can reach contradictory conclusions, and one of the conditions that permits such malleability of meta-analysis is the choice of

---

[10] For this latter reason I refrain from describing and illustrating the particular statistical analyses employed in tests of the inter-tool reliability of QATs, as I did in §7.2 on tests of the inter-rater reliability of QATs.

[11] This example was noted in Chapter 6. These QATS were the same that Moher et al. (1995) described. Jüni and his colleagues rightly noted that "most of these scoring systems lack a focused theoretical basis."

[12] (Moher, Jadad, and Tugwell, 1996).

quality assessment tool. The discordant results from the twenty-five meta-analyses performed by Jüni et al. (1999) are a case in point. This low inter-tool reliability also has philosophical consequences, which I explore in §7.4.

Such low inter-tool reliability might be less troubling if different quality assessment tools had distinct domains of application. The many biases present in medical research are pertinent to varying degrees depending on the details of particular circumstances (see Chapter 10), and so one might think that it is a mistake to expect that one quality assessment tool can apply to all circumstances. For some medical interventions, for instance, it is difficult or impossible to conceal the treatment from the experimental subjects and the investigators (surgical interventions, for example). No trial of such interventions will score well on a QAT that gives a large weight to allocation concealment. Such a quality assessment tool would be less sensitive to the presence or absence of sources of bias other than lack of allocation concealment, relative to tools that give little or no weight to allocation concealment. In such cases one might argue that since the absence of allocation concealment is fixed among the relevant studies, an appropriate QAT to use in this domain should not give any weight to allocation concealment, and one would only ask about the presence of those properties that might vary among the relevant studies.

On the other hand, one might argue that since we have principled reasons for thinking that the absence of allocation concealment can bias the results of a study, even among those studies that cannot possibly conceal subject allocation, an appropriate QAT to use in this case should evaluate the presence of allocation concealment (in which case all of the relevant studies would simply receive a zero score on allocation concealment), just as it ought to evaluate the presence of allocation concealment in a scenario in which the investigators can conceal subject allocation. The latter consideration should be generally compelling because we want to estimate the true effectiveness of medical interventions, and such estimates ought to take into account the full extent of the potential for biases in the relevant evidence, regardless of whether or not it was possible for the respective studies to avoid such biases.

Low inter-tool reliability would be less troubling if one could show that in principle there is only one good quality assessment tool for a given domain, or at least a small set of good ones which are similar to each other in important respects, because then one could dismiss the low inter-tool reliability as an artefact caused by the inclusion of poor quality assessment tools in addition to the good ones. Unfortunately, this is implausible. There are a plurality of equally fine QATs, designed for the same kinds of scenarios (typically: assessing the quality of randomized trials, which test the effectiveness of medical interventions).[13] Moreover, the empirical demonstrations of low inter-tool reliability involve the assessment of studies from a very narrow domain: for instance, the low inter-tool reliability shown by Jüni et al. (1999) involved assessing

---

[13] A review concluded that there were numerous quality assessment tools that all "represent acceptable approaches that could be used today without major modifications" (West et al., 2002).

studies of a single design (randomized trials) about a single hypothesis. Some of these tools are arguably inferior to others. But I argue below that, among the reasonably good quality assessment tools, we lack a theoretical basis for distinguishing some as better than others, and so we are stuck with a panoply of QATs that disagree widely about the quality of medical studies and thus the quality of the evidence generated from those studies.

One might agree with the view that there is no uniquely best QAT, but be tempted to think that this is due only to the fact that the quality of a study depends on particularities of the context (e.g., the particular kind of study in question and the form of the hypothesis being tested by that study). Different quality assessment tools might, according to this thought, be optimally suited to different contexts. While this latter point is no doubt true—above I noted that some are designed for assessing particular kinds of studies, and others are designed for assessing studies in particular domains of medicine—it does not explain their low inter-tool reliability. That is because, as above, the low inter-tool reliability is demonstrated in narrowly specified contexts.

Among medical scientists there is some debate about whether or not quality assessment tools ought to be employed at all.[14] The low inter-rater and inter-tool reliability of these tools might suggest that resistance to their use is warranted. There are three reasons, however, that justify the use of QATs. First, when performing a meta-analysis, a decision to not use an instrument to differentially weight the quality of primary-level studies is equivalent to weighting all the primary-level studies to an equal degree. So, whether one wishes to or not, when performing a meta-analysis one assigns weights to the primary-level studies, and the remaining question is simply how arbitrary one's method of weighting is. Assigning equal weights regardless of methodological quality is maximally arbitrary. The use of these tools to differentially assess studies is an attempt to minimize such arbitrariness. Second, as argued in §7.1, one must account for fine-grained methodological features in order to make a well-informed inference, and QATs help with this. Third, there is some empirical evidence that suggests that studies of lower quality have a tendency to overestimate the effectiveness of medical interventions, and so the use of quality assessment tools helps to estimate the true effectiveness of interventions. In short, despite their low inter-rater and inter-tool reliability, these tools are an important component of medical research.

7.4 Underdetermination of Evidential Significance

Quality assessment tools are used to estimate the quality of evidence from medical studies, and the primary use of such evidence is to estimate the effectiveness of medical interventions. These tools differ substantially in the weight assigned to various methodological properties (§7.1), and thus generate discordant estimates of evidential quality when applied to the same evidence (§7.3). The tools differ in arbitrary ways.

---

[14] See, for example, (Herbison, Hay-Smith, & Gillespie, 2006).

The low inter-tool reliability of QATs—together with the fundamentally arbitrary differences of their design—suggests that there is no uniquely correct value of the quality of evidence in this domain. This is an instance of a general problem I call the underdetermination of evidential significance.

Disagreement regarding the quality of evidence in particular scientific domains has been frequently documented with historical case studies, going back at least as far as Kuhn (1962). One virtue of examining the disagreement generated by the use of QATs is that such disagreements occur in highly controlled settings, are quantifiable using measures such as the κ statistic, and are about subjects of great importance. Such disagreements do not necessarily represent shortcoming on the part of the disagreeing scientists or the quality assessment tools they employ, nor do such disagreements necessarily suggest a crude relativism. Two scientists who disagree about the quality of a particular piece of evidence can both be rational because their differing assessments of the quality of the same evidence can be due to their different but equally reasonable weightings of fine-grained features of the methods that generated the evidence.

Kunda (1990) presented a summary of a research program in psychology about 'motivated reasoning,' a phenomenon in which subjects assess evidence differently depending on subjective idiosyncrasies. For example, after reading a scientific article that concludes that consuming caffeine is risky for females, female caffeine-consumers were less convinced by the article than were females who do not consume caffeine. In another study, subjects were presented with mixed evidence about the efficacy of capital punishment, and both supporters and opponents of capital punishment subsequently became more polarized in their respective views, which is perhaps best explained by a differential assessment of the mixed evidence. The same has been shown for assessments of evidence on the efficacy of policies for gun control.[15] These are classic examples of confirmation bias (see Chapter 10). In many instances of confirmation bias it is fair to think that the subjects involved in assessing the evidence are engaged in poor inductive reasoning. Motivated reasoning and confirmation bias, one might think, involve people who are poorly trained at evaluating evidence and who have been asked to perform complex intellectual tasks beyond their cognitive means, on politically charged topics or topics about which they are relatively unfamiliar.

But experiments demonstrating motivated reasoning have been performed on advanced graduate students and senior scientists, in which the scientists are asked to evaluate evidential quality from studies in their own discipline. Koehler (1993), for instance, found that evidence that confirmed scientists' prior beliefs was judged to be of higher quality than evidence that disconfirmed their prior beliefs, and the stronger such prior beliefs were, the stronger this effect was.[16] The phenomenon of motivated

[15] In an insightful discussion of these findings, Miller (2010) argues that the value-ladenness of science, together with the phenomenon of motivated reasoning, provides a novel argument for 'pragmatic encroachment' (the thesis that standards of knowledge depend in part on subjects' interests).

[16] Koehler also found that this effect was more pronounced when scientists were asked to evaluate coarse-grained features of evidence (like quality, relevance, and clarity of results) than when they were

reasoning is not limited to non-scientists assessing scientific evidence, but also occurs among scientists evaluating evidence in their own discipline.

Such findings might not surprise those who are versed in philosophy and sociology of science after Kuhn. But the distinctive feature of these findings, compared with much historically informed philosophy of science written in the wake of Kuhn, is that they are carefully controlled experiments. My discussion of quality assessment tools takes this one step further, showing disagreement about evaluations of the quality of evidence even when scientists have sophisticated and codified tools to measure the quality of evidence.

Concluding that there is no uniquely correct determination of the epistemic quality of a piece of evidence by appealing to the poor inter-rater and inter-tool reliability of QATs is not merely an argument from disagreement. If it were, then the standard objection to an argument from disagreement would simply note that the mere fact of disagreement about a particular subject does not imply that there is no correct or uniquely best view about the subject of disagreement. Although different quality assessment tools disagree about the quality of evidence from a trial, this does not imply that there is no true or best view regarding the quality of evidence from the trial—goes the objection—since the best tools might agree with each other about the evidence from the trial, and even more ambitiously, agreement or disagreement among these tools would be irrelevant if we just took into account the quality assessment of a trial by the uniquely best tool. The burden that this objection faces is the identification of the single best quality assessment tool or at least the set of good ones (and then hope that the set of good ones will have high inter-tool reliability). As noted in §7.3, medical scientists claim that there is simply a plurality of these tools that differ from each other in arbitrary respects.

More fundamentally, we lack a theory of scientific inference that would allow one to referee between the most sophisticated quality assessment tools. Recall the different weightings of the particular methodological features assessed by these tools, noted in Table 7.2. Another way to state the burden of the 'mere argument by disagreement' objection is that to identify the best quality assessment tools, one would have to possess a principled justification for determining the optimal weights of the methodological features included in the tools. That we do not presently have such a principled justification is an understatement.

Consider this compelling illustration of the arbitrariness involved in the assignment of weights to methodological features in quality assessment tools. Cho and Bero (1994) employed three different algorithms for weighting methodological features. Then they tested the three weighting algorithms for their effect on quality scores of trials, and their effect on the inter-rater reliability of such scores. They selected for further use—with no principled basis—the weighting algorithm that had the highest inter-rater

---

asked to evaluate fine-grained features of evidence (like the adequacy of randomization procedures)—which supports the use of fine-grained evidence assessment tools such as QATs.

reliability. Cho and Bero explicitly admitted that nothing beyond the higher inter-rater reliability warranted the choice of this weighting algorithm.[17] Medical scientists have no foundation for developing a uniquely good QAT, and so they resort to a relatively arbitrary basis for their development.

One could press this objection by noting that while it is true that we presently lack an inductive theory that could justify a unique system for weighting the various methodological features, it is overly pessimistic to think that we cannot have a principled basis for identifying a uniquely best weighting system. However, as the work of Cho and Bero suggests, the poor inter-tool reliability noted in §7.3 is due to arbitrary differences in how methodological features are weighed. One could insist on claiming that it is possible that in the future we will develop a theory of inference that would allow us to identify a uniquely best weighting system. There is a point at which one cannot argue against philosophical optimism.

One could put aside the aim of finding a principled basis for selecting among the available QATs, and instead perform a selection based on their historical performance. Since these tools are employed to estimate the quality of evidence from medical studies, and such evidence is used for supporting hypotheses regarding the effectiveness of medical interventions, the historical approach would involve selecting QATs based on the fit between (i) the degree to which such hypotheses are presently confirmed and (ii) the quality of the evidence for such hypotheses available at a particular time in the past as determined in retrospect by current QATs. The best quality assessment tool would be the one with the highest average fit between (i) and (ii). Unfortunately, such an assessment of these tools would be limited by a fundamental epistemic circularity. The degree to which hypotheses regarding effectiveness of medical interventions are confirmed is based on the total available evidence, summarized by a careful systematic review (which usually takes the form of a meta-analysis), appropriately weighted to take into account relevant methodological features of those studies. But of course, those very weightings are generated by QATs. The historical approach to assessing these tools, then, itself requires the use of these tools.

The underdetermination of evidential significance is not the same problem that is associated with Duhem and Quine. One formulation of the standard underdetermination problem—underdetermination of theory by evidence—holds that there are multiple theories compatible with a given body of evidence. The underdetermination of evidential significance is the prior problem of settling on the quality of evidence in the first place. Perhaps an appropriate term for the present problem is just the inverse of the Quinean locution: *underdetermination of evidence by theory*. Quality of evidence is underdetermined by theories of scientific methodology.

---

[17] They rightfully claimed that such arbitrariness was justified because "there is little empiric evidence on the relative importance of the individual quality criteria to the control of systematic bias [sic]." We lack reasons to prefer one weighting of methodological features over another, regardless of whether one thinks that such reasons are purely empirical or also include principled considerations.

7.5 Conclusion

There is a plurality of methodological features, such as randomization and subject allocation concealment, that must be considered when assessing evidence in medical research, and there are numerous ways to do so. Quality assessment tools are codified and relatively sophisticated methods of assessing medical evidence. However, these tools vary in their composition, and their inter-rater reliability and inter-tool reliability is low. This, in turn, is a compelling illustration of a more general problem: the under-determination of evidential significance. Disagreements about the quality of evidence are ubiquitous in science. Such disagreement is especially striking, however, when it results from the employment of carefully codified tools designed to quantitatively assess quality of evidence. The tools discussed in this chapter are the best instruments available to medical scientists to assess quality of evidence from clinical research, yet when applied to what is purported to be the best-quality evidence in medicine (namely, evidence from randomized trials), different users of the same QAT, and different QATs applied to the same evidence, lead to widely discordant assessments of the quality of evidence. This is a fundamental form of malleability in medical research.

8

Measuring Effectiveness

8.1 Introduction

Clinical research is performed to estimate the effectiveness of medical interventions. In this chapter I argue that there are three widespread problems in measuring the effectiveness of medical interventions: the use of poor measuring instruments, the use of misleading analytic measures, and the assumption that measurements in an experimental setting are sufficient to infer a general capacity of effectiveness. Each of these problems contributes to overestimating the effectiveness of medical interventions. The problems suggest corrective principles—medical research should use appropriate measuring instruments, truth-conducive analytic measures, and reliable methods of extrapolation. The application of such principles would generally lead to lower—yet more accurate—estimates of the effectiveness of medical interventions than is presently the case.

By far the most common method for measuring effectiveness of interventions is the randomized trial.[1] A randomized trial involves administering an experimental intervention to one group of subjects (the experimental group), administering a placebo or competitor intervention to another group of subjects (the control group), measuring parameters of the subjects, comparing the values of those parameters between the two groups, and if the values of parameters differ between groups, inferring that the intervention has a general capacity to cause that difference. Trials usually have methodological safeguards to minimize systematic error, prominently including the random allocation of subjects to groups, and concealment of the group assignment from both the investigators and the subjects (in Chapter 10 and elsewhere throughout this book I note ways in which such safeguards often fail). These details aside, the measurement of effectiveness involves three steps: the use of a measuring instrument (or a measuring technique more generally), the analysis of measured values, and the extrapolation of analyzed values to a target population.

As argued in Chapters 2 and 3, effectiveness of medical interventions is a capacity to improve the health of patients by targeting disease. This is not an intrinsic causal

---

[1] As I argue in this chapter and elsewhere throughout this book, the exclusion of evidence from other kinds of methods in the measurement of effectiveness is a significant epistemic limitation. But since this reliance on clinical trials is so ubiquitous, my focus in this chapter is on this method.

capacity; effectiveness is a relational property in which the relata are a causal capacity of the intervention and properties of people with a particular disease. The properties that must be modulated by a medical intervention in order for that intervention to be effective are either the constitutive causal basis of a disease or symptoms of a disease that cause harm to those with that disease. In Chapter 2 I call these two individually sufficient conditions for effectiveness CAUSAL TARGET OF EFFECTIVENESS and NORMA-TIVE TARGET OF EFFECTIVENESS. In Chapters 2 and 3 my aim is to articulate a view of what effectiveness is (a conceptual and metaphysical question), whereas in the present chapter my aim is to articulate methodological problems associated with how we measure effectiveness.

For any measurement one needs a measuring instrument. In clinical practice and medical research many instruments are employed to measure various kinds of param-eters, including subjective patient-reported parameters (such as reports of well-being), physician-reported parameters (such as appearance of lethargy), institutional param-eters (such as number of days in an intensive care unit), and physiological parameters (such as blood sugar concentrations). For example, the Hamilton Rating Scale for Depression (discussed below) measures several of these kinds of outcomes, including a patient's report of sadness and quality of sleep, a physician's assessment of the patient's fidgetiness, and physiological correlates of anxiety. Sometimes the outcome of interest in a clinical trial is simple, like an event such as death, in which case the appropriate measuring instrument is whatever is required to determine that the event has occurred. I use the term 'instrument' very broadly to include any tool or technique employed to estimate values of measurands. In §8.2 I describe examples of measuring instruments in medical research, and argue that many such instruments are unreliable tools for measuring effectiveness, because they are not specific to properties relevant to infer-ring effectiveness of interventions.

Once parameters are chosen and instruments have been employed to measure val-ues of those parameters among subjects in a trial, those values must be interpreted in some way to assess whether, and if so to what extent, an intervention modifies the val-ues of those parameters. Several analytic tools are widely employed in medical science as measures of effectiveness; these are called outcome measures, and the numerical outputs of outcome measures are called effect sizes. In §8.3 I describe several outcome measures and argue that the most widely employed class of outcome measures is mis-leading. From the perspective of a patient or a physician who is deciding whether or not to use or prescribe a particular treatment, the best outcome measures are absolute measures, which, unlike relative measures, take into account the baseline value of the parameter being measured.

The aim of measuring the effectiveness of medical interventions is to aid in deci-sions regarding treatment, which involves predicting outcomes in target patient popu-lations (see §8.4). One method for making such predictions is simple extrapolation from the quantitative results of trials to a target population. Simple extrapolation is often implicitly employed in medical research and policy, and is sometimes explicitly defended as a method for extrapolation. But I argue that simple extrapolation is unreliable, and it tends to overestimate the effectiveness of medical interventions.

Thus, clinical research involves a chain of measurands, in which the value of one measurand is used to infer the value of the next measurand in the chain. This is typical for the epistemology of measurement: measuring the temperature in my backyard involves measuring the height of mercury in a glass tube; measuring the rate of expansion of the universe involves measuring Hubble's Constant by measuring wavelengths of light undergoing redshift.[2] The ultimate measurand of interest in medical research is the effectiveness of a medical intervention. Estimating this measurand is based (at least in part) on a prior measurand: the capacity of the medical intervention, in a controlled experimental setting, to cause a difference in the value of the parameter of interest between the experimental group and the control group. This in turn involves measurement of the value of that very parameter in those subjects. At each of the three links of this chain of measurands there are challenges that are often not adequately resolved in medical research.

In short, the measurement of effectiveness of medical interventions faces three methodological challenges, associated with the choice of measuring instrument (§8.2), outcome measure (§8.3), and method of extrapolation (§8.4). I am not the first to note these challenges. But in this chapter I argue that in practice these challenges contribute to overestimating the effectiveness of medical interventions. These problems render current regulatory standards too low (§8.5). If these challenges were better addressed, estimates of the effectiveness of medical interventions would be more accurate, and lower than they are now, and fewer medical interventions would be approved for clinical use.

The measurement of effectiveness of medical interventions can be modeled in a variety of ways—in Chapter 11 I employ a formal device to model the measurement of effectiveness, in which I present the master argument for medical nihilism, and in Appendix 5 I suggest an alternative way to model the measurement of effectiveness. Both approaches entail that the effectiveness of medical interventions is now exaggerated.

8.2 Instruments

To determine the values of parameters of subjects in the experimental and control groups of a trial, one needs a measuring instrument. Such instruments can vary in a number of respects. Instruments can be simple, particularly when the measurand is an event (such as death), or they can be multifaceted, particularly when the measurand is characterized by medical constructs (such as depression). Another dimension of instruments is their inferential directness: some instruments involve relatively direct

---

[2] For recent work on the epistemology of measurement, see (Chang, 2004), (Alexandrova, 2008), (van Fraassen, 2008), (Tal, 2011), (Teller, 2013), and (Tal, 2016).

measures of properties of interest, in that the value determined by the instrument requires only a few (usually reliable) inferences to determine the value of the measurand (such as blood sugar concentration). Other instruments are inferentially indirect, in that they are measures of a proxy of the measurand of interest, and the measurement procedure requires more inferences (which are often less reliable) from the value of the measured parameter to the value of the measurand of interest (such as the measurement of cholesterol as an indicator of heart disease). As with all measuring instruments, two desiderata are sensitivity and specificity: a measuring instrument should be sensitive to the true values of the measurand of interest, and should be specific to those values (that is, sensitive *only* to the true values of the measurand of interest). The employment of certain instruments, some of which are widely used in medical research, contributes to overestimations of the effectiveness of interventions.

Here is an example of an inferentially indirect instrument that is nonspecific to values of the measurand of interest. Some evidence suggests that certain interventions can reduce the 'white lesions' that are physiological correlates ('biomarkers') of multiple sclerosis. White lesions are the result of the demyelination of the sheaths that surround the axons of neurons, and are not themselves the causes of multiple sclerosis (which remain unknown). The hope is that if white lesions are a proximal cause of multiple sclerosis symptoms (below I note that this is doubtful), then mitigating white lesions will mitigate symptoms. Some trials have suggested that some interventions have the capacity to decrease the number of white lesions. The instruments in such trials measure a proxy of the measurand of interest: white lesions as a surrogate of patient-level symptoms that constitute the harms to patients with multiple sclerosis.[3] Under the assumption that mitigating white lesions will mitigate patient-level symptoms, the results of this measurement license an inference that the drugs under investigation are effective at mitigating multiple sclerosis symptoms. Unfortunately, evidence suggests that such drugs have little impact on symptoms. The use of white lesions as a proxy for patient-level symptoms is nonspecific, because it is sensitive to values of parameters that are only weakly correlated with the measurand of interest. The inferential assumption noted above is probably false: demyelination of axon sheaths is probably a common cause of white lesions and patient-level symptoms, and since the direction of causal relevance is not (as far as we can tell) from white lesions to patient-level symptoms, measuring a reduction in white lesions does not warrant an inference about the reduction of patient-level symptoms. The measuring instrument in this case overestimates the true value of the measurand of interest.

Here is an example of a multifaceted measuring instrument: the Hamilton Depression Rating Scale (HAMD), which is one of the most commonly employed instruments in trials testing the effectiveness of antidepressants. It is a questionnaire composed of seventeen questions, each of which has between three to five possible

---

[3] See (Lavery et al., 2014) for a discussion of various outcomes measured in trials of interventions for multiple sclerosis.

answers with a corresponding numerical score, and the sum of these scores is interpreted as measuring the severity of depression.[4] The total score can range from 0 to 52 points, and scores are interpreted in terms of severity of depression as follows: 0–7: normal; 8–13: mild; 14–18: moderate; 19–22: severe; ≥ 23: very severe. These scores are determined for subjects in a trial, and if an antidepressant is effective one expects to observe a greater decrease in average score for subjects in the experimental group compared with subjects in the control group. However, this scale is a nonspecific instrument with regard to the measurand of interest, namely, intensity of depression. That is because many of the questions included on the HAMD are mostly irrelevant to this measurand.

Some of these questions probe core elements of depression, albeit at a coarse grain. For instance, the question on 'suicidality' is scored as follows: "0 = Absent. 1 = Feels life is not worth living. 2 = Wishes he were dead or any thoughts of possible death to self. 3 = Suicidal ideas or gesture. 4 = Attempts at suicide (any serious attempt rates 4)." Thus, the greater the degree of suicidality of a subject, the greater the HAMD score.[5] Other questions in the scale probe features of a person's state that are less central to depression. For instance, there are three questions regarding insomnia, corresponding to three phases of night (early, mid, and late), and there are six possible points for these questions. There are four possible points associated with fidgeting. Thus, if an intervention causes people to sleep better and fidget less, the corresponding HAMD reduction could be up to ten points. To put this in perspective, some guidelines have held that a reduction of three points on this scale entails that an intervention is an effective antidepressant. A small improvement in sleep or a decrease in fidgeting caused by an intervention would warrant approval as an effective antidepressant, despite the fact that the intervention might not mitigate any of the fundamental symptoms of depression, such as low mood, anhedonia, and feelings of worthlessness, guilt, and hopelessness. This scale is an example of a multifaceted instrument which overestimates effectiveness of interventions because it is not sufficiently specific.[6]

A curious HAMD question is titled 'Insight,' and is scored as follows: "0 = Acknowledges being depressed and ill. 1 = Acknowledges illness but attributes causes to bad food, climate, overwork, virus, need for rest, etc. 2 = Denies being ill at all." Thus, a tired patient who suspects that her illness is caused by a gluten allergy automatically gets an extra point on her HAMD score. Philosophical critics and rugged cowboys,

---

[4] There are various versions of the HAMD, but here I describe the original proposed by Hamilton (1960). There are an additional four questions which do not contribute to the score. This instrument has been hugely influential in clinical psychiatric research. Note that sometimes in the clinical literature alternative abbreviations are used for the scale.

[5] In Chapter 9 I discuss the role that measuring instruments play in the underestimation of the harms of medical interventions—this question on suicidality is an example there.

[6] One might hold that among patients with depression, their insomnia is likely to be caused by their disease, and thus if a drug improves a depressed person's sleep then either the drug is (i) intervening in the pathophysiology of depression, or at least is (ii) offering relief of symptoms of depression. (i) is excessively optimistic, but (ii) seems reasonable. Such a drug would be a soporific rather than an antidepressant.

beware! Denying one's alleged depression earns you *two* extra points. This question highlights a serious measurement problem: if an experimental intervention causes a subject who initially denies being ill to later claim that they are ill—say, because the intervention itself causes illness—then that subject's HAMD score would go *down* by two points, thereby making the intervention appear to be an effective antidepressant, despite the fact that it caused the symptoms.

The use of instruments for measuring the effectiveness of interventions should be sensitive to the temporal dimension of the disease being treated (the term of art in medicine is the 'clinical course' of the disease). Unfortunately, instruments are often used in ways that are insensitive to the clinical course of a disease. For instance, when researchers were testing high-dose chemotherapy for breast cancer, they assessed presence of cancer after eighteen months of treatment (see (Brownlee, 2008)). This temporal range was adopted from research on blood cancers, in which high-dose chemotherapy is effective. After eighteen months it appeared that the high-dose chemotherapy had prevented recurrence of breast cancers. However, breast cancers grow slower than blood cancers, and so eighteen months was an inappropriately short duration to measure the outcome of the therapy. Later studies that used a longer duration found that high-dose chemotherapy did more harm than good for breast cancers. The difference in growth rates between cancer types explains why high-dose chemotherapy is more effective in blood cancers than in breast cancers: since chemotherapeutic drugs operate by interfering with mechanisms of cell division, cells that divide rapidly are more susceptible to chemotherapy (and conversely, slower-growing tumors are less susceptible to chemotherapy).[7] The initial studies that suggested that high-dose chemotherapy is effective for breast cancer employed an instrument in a way that was not sufficiently sensitive to the temporality of the disease.

Another example in which an inappropriate temporal range has been used are trials testing the effectiveness of methylphenidate (Ritalin) to treat attention deficit hyperactivity disorder (ADHD). Many of these trials have only lasted a few weeks, and meta-analyses of these trials suggest that methylphenidate has a small but positive effect on ADHD symptoms in the short term. But in studies that follow-up with patients from the longest trial performed thus far, there is no beneficial difference in ADHD symptoms among children who had been on methylphenidate compared to children who had not. There are, however, apparent harms caused by methylphenidate in the long run, including a decrease in body height and mass.[8] Thus an inappropriately short temporal range of most studies on the effectiveness of methylphenidate contributes to

---

[7] This example provides support to the view that mechanistic knowledge should be taken into account when assessing effectiveness (see §8.4).

[8] These follow-ups have been done at three years and eight years after the trial, which was the fourteen-month MTA trial sponsored by the National Institute of Mental Health (NIMH) (discussed in more detail in Chapter 9). For the empirical findings mentioned here, see (Schachter, Pham, King, Langford, & Moher, 2001), (Jensen et al., 2007), and (Molina et al., 2009).

an overestimation of the effectiveness of the drug (and an underestimation of its harm profile).

The use of indirect instruments contributes to overestimation of effectiveness, since the causal link between the measurand used in indirect instruments and the measurand of interest is often not tight or well understood. For example, high cholesterol levels are thought to be a cause of heart disease, and so to avoid heart disease, cholesterol-lowering drugs are prescribed to many people. Trials have shown that these drugs are effective at lowering cholesterol. With respect to the end of avoiding heart disease, many trials employ an indirect instrument (measurement of cholesterol levels), and under the assumption that high cholesterol causes heart disease, the evidence from these trials warrants an inference that cholesterol-lowering drugs are effective at mitigating heart disease. Unfortunately, trials that employ the more direct instrument of measuring heart disease have shown that these drugs are barely effective at mitigating heart disease.[9] To use the analysis from Chapter 2, the problem with using indirect instruments to measure surrogate outcomes is that an intervention that modulates a surrogate outcome does not satisfy CAUSAL TARGET OF EFFECTIVENESS, because the surrogate outcome is not the constitutive causal basis of the disease, and similarly, an intervention that modulates a surrogate outcome does not satisfy NORMATIVE TARGET OF EFFECTIVENESS, because by definition surrogates are stand-ins for the patient-level outcomes that matter (the harms of the disease).

In principle, the measurement bias of instruments in clinical research is symmetric with respect to estimating effectiveness: the above problems could lead to underestimation of effectiveness as often as they lead to overestimation. For example, trials on antidepressants could employ an instrument that is relatively insensitive to changes in a person's core symptoms of depression, in which case such trials would tend to underestimate the effectiveness of experimental interventions. Trial designers (I assume) are often aware of the details regarding their measuring instruments, which generate a trade-off between the predilection of a trial to overestimate effectiveness and the predilection of a trial to underestimate effectiveness. Moreover, trial designers have strong incentive to err on the side of overestimating rather than underestimating effectiveness—this is a contingent sociological fact based on pressure to publish among academic scientists and pressure to develop profitable products among corporate scientists. Since trial designers have strong incentive to err on the side of overestimating effectiveness, they tend to do so, as illustrated by the above examples.

McClimans (2010) notes that there are thousands of measuring instruments in medicine, yet the theoretical underpinning of such instruments is often poorly understood, and researchers and regulators often ignore the consequences of this. I have

---

[9] See (Moynihan & Cassels, 2005). The use of surrogate outcomes as measurands in clinical research is ubiquitous. An example of a medical intervention that I return to in detail in Chapter 9 is rosiglitazone (Avandia), which was approved by the FDA based on its capacity to modulate glycemic levels, which is a surrogate outcome.

MEASURING EFFECTIVENESS   119

argued here that problems of measuring instruments in clinical research contribute to an overestimation of the effectiveness of medical interventions.

Once evidence is gathered with such measuring instruments, the evidence is often analyzed and presented in such a way as to make the experimental interventions appear more effective than they are—a problem I now turn to.

8.3 Measures

Many outcome measures are employed in clinical research. An outcome measure is an abstract formal statement describing a relation between the value of a measurand in the control group and the value of that measurand in the experimental group. When substantive values for such measurands are substituted into an outcome measure, the result is a quantitative estimation of efficacy—the strength of an alleged causal relation manifest in the trial—and this quantity is called an effect size.

If the measured parameters are continuous (such as blood sugar concentration), a common outcome measure is the standardized mean difference (SMD):

$$SMD = (\mu_1 - \mu_2) / \sigma$$

where μ_1 is the mean value of a parameter of interest for the experimental group, μ_2 is the mean value of the same parameter for the control group, and σ is a measure of the variance of the value of the parameter (for some statistics σ is measured in the control group and for other statistics the variance from both groups is pooled to determine σ). A ubiquitous practice in medical research is to use SMD as the basis of more complicated analytic statistics. The simple measure $\mu_1 - \mu_2$ is important because it measures the absolute difference between mean values of the parameter of interest, which is an estimate of the difference-making capacity of the intervention.

For both continuous and dichotomous parameters, the choice of outcome measure is important and can have significant influence on the estimation of effectiveness. In what follows I focus on outcome measures for dichotomous parameters. There are two main classes of outcome measures for dichotomous parameters: absolute measures and relative measures. I argue below that absolute outcome measures are superior to relative outcome measures.

For dichotomous parameters (such as death), outcome measures include odds ratio, relative risk (sometimes called risk ratio), relative risk reduction, risk difference (sometimes called absolute risk reduction), and number needed to treat. To define these, one constructs a two-by-two table (this is the same as Table 6.1 in Chapter 6, repeated here as Table 8.1 for ease of reference). Consider a study that has an experimental group (E) composed of subjects who receive the experimental intervention, and a control group (C) composed of subjects who do not receive the experimental intervention (perhaps they receive a placebo), in which a binary outcome is measured as present (Y) or absent (N), where the number of subjects with each outcome in each group is represented by letters (a–d), as follows.

Table 8.1. A 2 × 2 table for defining outcome measures for a dichotomous parameter

	Group	Outcome
	Y	N
E	a	b
C	c	d

Relative risk (RR) is defined as:

$$RR = \left[a/(a+b)\right]/\left[c/(c+d)\right]$$

Relative risk reduction (RRR) is defined as:

$$RRR = \left[\left[a/(a+b)\right]-\left[c/(c+d)\right]\right]/\left[c/(c+d)\right]$$

Risk difference (RD) is defined as:

$$RD = a/(a+b) - c/(c+d)$$

Number needed to treat (NNT) is defined as:

$$NNT = 1/\left[\left[a/(a+b)\right]-\left[c/(c+d)\right]\right]$$

It also can be useful to define these outcome measures in terms of conditional probabilities. The probability of a subject having a Y outcome given that the subject is in group E, $P(Y|E)$, is $a/(a+b)$, and likewise, the probability of having a Y outcome given that the subject is in group C, $P(Y|C)$, is $c/(c+d)$. Thus, for example, we have:

$$RR = P(Y|E)/P(Y|C)$$

and

$$RD = P(Y|E) - P(Y|C)$$

A widespread and misguided practice is to report RR or RRR but not RD or NNT. Broadbent (2013) dubs the overreliance on relative outcome measures in epidemiology 'risk relativism' (he canvasses several alleged justifications for the widespread use of relative measures like RR and finds them all wanting). My concern here is with a nefarious consequence of risk relativism. Both physicians and patients overestimate the effectiveness of interventions when presented with relative measures, and their estimates are more accurate when they are presented with both relative and absolute measures or with absolute measures alone. This finding has been replicated many times in different contexts.[10] This overestimation of effectiveness occurs because the use of relative measures, such as RR or RRR, promotes the base rate fallacy.

[10] See, as examples: (Nexøe, Gyrd-Hansen, Kragstrup, Kristiansen, & Nielsen, 2002), (Forrow, Taylor, & Arnold, 1992), (Naylor, Chen, & Strauss, 1992), (Sorensen, Gyrd-Hansen, Kristiansen, Nexoe, & Nielsen, 2008), (Bobbio, Demichelis, & Giusetto, 1994), and (Malenka, Baron, Johansen, Wahrenberger, & Ross, 1993).

The base rate fallacy involves making an inference about the probability of an event without taking into account the prior probability of that event. For example, in Appendix 1, if someone estimated the probability that a person with a positive disease test had the disease without taking into account the background (prior) probability of that disease, then they would be committing the base rate fallacy. Psychologists have found that people are especially prone to committing the base rate fallacy. The base-rate fallacy can lead people far astray in their inferences. This is the case with inferring effectiveness of medical interventions with relative measures like RR, as I argue below.

A question of central concern for a patient is: to what extent would using this particular intervention change the probability of me having the outcome in question? Two epistemological notions are pertinent here: *change* and *probability*. Of course, the outcome could have occurred had the patient not used the intervention. So the patient's question is counterfactual: to what extent would using this particular intervention change the probability of me having the outcome from the probability of me having had the outcome had I not used the intervention? An equivalent way to put the patient's question is: by how much does the probability of Y change if I consume E instead of C? That is represented by $P(Y|E) - P(Y|C)$. And that is the risk difference.

Because the risk difference is a difference between posterior probabilities, the risk difference is a measure of the probability of experiencing Y that takes into account the prior probability of Y—prior probabilities are already incorporated in posterior probabilities (see Appendix 3). So, inferring the effectiveness of an intervention based on RD does not necessarily lead one to commit the base rate fallacy (though it is still possible for one to commit the base rate fallacy in other ways when estimating the effectiveness of interventions). A ratio of posterior probabilities (RR) does not have this property, because the prior probability of Y is part of both posterior probabilities of the ratio (RR), and so those prior probabilities cancel out. Thus the prior probability is not accounted for in RR. So, inferring effectiveness based on RR contributes to committing the base rate fallacy (Appendix 3). This explains why physicians and patients overestimate effectiveness when presented with relative effect sizes.

Another way to represent the patient's question is to multiply two factors: a factor that represents the difference-making capacity of the intervention, and a factor that represents the baseline probability of the outcome in question. In Appendix 3 I prove that RD represents both important epistemological notions and thus is a reliable outcome measure for making inferences about effectiveness, and I show that RR is not sensitive to the baseline (prior) probability of the outcome in question, and thus facilitates the base rate fallacy.

Effect sizes are derived from controlled experimental settings, and in §8.4 I argue that care should be applied when extrapolating from the results of a trial to making an inference about how beneficial an intervention will be for a particular patient outside the controlled setting of a trial. Nevertheless, to address the patient's central question articulated above, we need a measure that represents the capacity of an intervention to change the probability of the beneficial outcome in question. RR does not do this, while RD does.

Here is a related, decision-theoretic argument in favor of RD over RR. Worrall (2010) rightly notes that the choice of using a medical intervention is a decision that ought to be modeled according to our best normative theory of decision-making. I will formulate this insight. Let x be the intended beneficial effect of a drug—say, avoiding a heart attack—which brings utility $U(x)$ to a patient, and let the harmful effects of the drug be y_i, which brings utility $U(y_i)$ to a patient (these are negative). Decision theory holds that in standard cases one should take that action if it would bring more utility than not taking that action, and if it would not, then do not. Of course, any of x and y_i could have occurred without using the drug. Thus the expected utility (EU_D) of using the drug (D), compared with not using the drug (~D), is:

$$EU_D = \left[P(x|D) - P(x|\sim D) \right] U(x) + \forall i \Sigma \left[P(y_i|D) - P(y_i|\sim D) \right] U(y_i)$$

Note that the leftmost multiplicand in the leftmost term of EU_D is best estimated by RD. Indeed, a naïve estimation of $P(x|D)$ would just be based on the frequencies a/(a+b), and a naïve estimation of $P(x|\sim D)$ would just be based on the frequencies c/(c+d), and since the difference between the former and the latter is just RD, a naïve estimation of the leftmost multiplicand in the leftmost term of EU_D would simply be based on RD. I call this naïve for reasons described in §8.4. Nevertheless, RD is an estimator of this term required for the expected utility calculation—relative measures are simply not an option for this estimation. In Appendix 4 I prove that RD is *EU-sufficient* (for given costs and utilities of two interventions, RD is sufficient to compare the expected utilities of those interventions), and further prove that RR is *EU-insufficient* (for given costs and utilities of two interventions, RR is *not* sufficient to compare the expected utilities of those interventions). This finding provides strong reason for requiring the use of RD in measuring the effectiveness of medical interventions.[11]

To illustrate the problem that arises when not taking P(Y) into account with relative measures of effectiveness, consider the drug alendronate (Fosamax), claimed to allegedly cause an increase in bone density in women, used with the aim of decreasing the frequency of bone fractures. A large trial compared the drug to placebo over a four-year period.[12] The evidence from the trial was touted as showing that the drug reduces the risk of hip fractures by 50 percent—this was a relative risk reduction (RRR). However, only 2 percent of the women in the control group had hip fractures during the four years of the trial, while 1 percent of the women in the experimental group had hip fractures. Thus the RD was a mere 1 percent—the absolute difference in hip fracture rates between the experimental group and the control group, that is RD, was only 1 percent—after consuming the drug for four years.

---

[11] One ought not assume that the expected utility of using a medical intervention is positive, given only a large relative or absolute effect size.

[12] The trial was reported in (Black et al., 1996). See (Moynihan & Cassels, 2005) for commentary. It was only women at high risk of hip fractures—those who had already had hip fractures—who were included as subjects, and thus the trial subjects were not representative of the broader population (which raises the problem of extrapolation, to which I turn in the following section).

You might maintain the perplexed question: how effective is alendronate sodium effective? After all, we have two outcome measures reporting two effect sizes:

$$RRR = 50\%$$
$$RD = 1\%$$

So, does alendronate sodium decrease the chance of hip fractures in the relevant population by 50 percent or 1 percent? The answer is that it does both, because the question is ambiguous. For a particular patient, the probability of having a hip fracture after taking alendronate sodium decreases from 2 percent to 1 percent, and so, since 2 – 1 = 1, the chance of having a hip fracture decreases by 1 percent. But since 1 (of anything) is 50 percent of 2 (of anything), the probability of having a hip fracture after taking alendronate sodium decreases by 50 percent. Which effect size should a particular patient and her physician be impressed by? The absolute measure is necessary to make a reliable inference about effectiveness and an informed treatment decision. The probability of having a hip fracture after taking alendronate sodium decreases by 1 percent. Alendronate is barely effective, even in the most at-risk patients. The use of a relative outcome measure makes the drug seem much more effective than it in fact is.

Here is another example. The Helsinki Heart Study tested the capacity of gemfibrozil to decrease cholesterol levels and thereby decrease cardiac disease and death. After five years of taking the drug, the subjects in the experimental group had a reduced relative risk of cardiac disease of 34 percent, but since the baseline rate of cardiac disease is so low, this amounted to an absolute risk difference (RD) of only 1.4 percent.[13]

The reliance on relative outcome measures at the expense of absolute outcome measures is ubiquitous. One empirical finding estimates that about three quarters of clinical trials only report relative outcome measures.[14] This, together with the fact that people overestimate the effectiveness of medical interventions when provided with relative outcome measures, entails that on average people overestimate the effectiveness of interventions. Effectiveness always should be measured and reported in absolute terms (using measures such as RD). This would have the result that estimates of the effectiveness of medical interventions would be more accurately deemed lower than they now are.

One might object that a medical intervention with a low absolute effect size could nevertheless be considered effective, because if the medical intervention were used by a large number of people, then many people would experience the beneficial outcome of the intervention. This is especially the case with those medical interventions that are widely used today as preventive medications, such as drugs for lowering cholesterol and blood pressure. For example, if a cholesterol-lowering drug has a 1 percent absolute

---

[13] The trial was reported in (Frick et al., 1987). Notably, there was no difference between the groups in the death rate.

[14] Many commentators have noted that risk relativism is widespread. King, Harper, and Young (2012) analyzed a sample of articles published in a number of medical and epidemiology journals and found that 75 percent of articles reported only relative measures.

reduction in the risk of death, and ten million people consume the drug, then 100,000 lives are saved. That is a great outcome. However, it is not obviously great from the perspective of a particular typical patient—as noted above, a particular patient must take into account much more information than the risk difference for an outcome when deciding if the intervention is the right thing for them. Nevertheless, despite its low absolute effect size, a patient may still consider it worthwhile.

One rationale for the use of low-RD interventions could be similar to the rationale for the use of vaccines: most people who are vaccinated against a disease would not have developed the disease had they not been vaccinated, and thus they do not directly receive a benefit from the intervention, but the widespread use of vaccines is nonetheless warranted because the practice decreases the overall number of people who develop the disease (thanks to 'herd immunity'). This way of conceiving of the benefits of vaccines requires thinking of the benefit accrued to a population rather than any particular individual. However, that is not how most drugs with low effect sizes—drugs that lower cholesterol, say—should be thought of. An individual patient and her physician want to know that if they use a particular intervention then there is a reasonably good chance that the intervention will be effective for this particular patient. For drugs with low absolute effect sizes like the ones I have been discussing above, that is not the case. More to the point, as noted above, the decision to take a drug is just that—a decision—and should be guided by our best framework for making decisions, which requires estimating the probabilities and utilities of the various possible outcomes of a decision. RD gives us such an estimate (the probability that the patient will experience the beneficial outcome of interest), but RR does not.

An objection related to the one above holds that interventions with high relative effect sizes but low absolute effect sizes are indeed effective—alendronate sodium cuts one's risk of hip fractures in half, after all—it is just that there are relatively few people for whom the intervention can be effective, because the baseline probability of a woman having a hip fracture is so low. If a woman were among the 2 percent who were going to have a hip fracture, then alendronate would cut that woman's risk in half, which (this objection goes) is significant. The trouble with such a response is that one cannot tell in advance if one is in the class of people for whom an intervention might be effective—namely, the class of people who would experience the negative outcome in question. From a particular patient's perspective—one who does not know in advance if she will have a hip fracture, say—a drug like alendronate sodium decreases her chance of having a hip fracture by a tiny amount. Another way of putting the point is: for a particular patient, an intervention with a low absolute effect size is very unlikely to provide any benefit at all.

To see this, consider the absolute outcome measure 'number needed to treat' (NNT). This is an intuitive measure: it tells you how many people would have to use an intervention in order to achieve one of the outcomes of interest. NNT is just the inverse of RD. For example, if an intervention has an RD of 1 percent, then the NNT is 100. That is, one hundred people would need to use the intervention in order to achieve one positive outcome. In other words, only one of the 100 people who used the intervention would experience the beneficial outcome, while the other 99 would not. The outcome of interest would not be experienced by the vast majority of the people that consume the drug. As above, when deciding whether or not to use an intervention, a patient or physician wants to know to what extent would using this intervention change the probability of the outcome in question. To determine this, measures like RD and NNT are needed.

As the above examples illustrate, when the base rate of an outcome is low, an intervention employed to avoid that outcome could have a large relative effect size but a small absolute effect size. Schwartz and Meslin (2008) suggest that the use of absolute measures could cause patients to forgo treatment in cases similar to those above (in which the absolute effect sizes are tiny), and they suggest that this is a reason in favor of the use of relative measures. Their argument is: for a patient to make an autonomous medical decision they must be informed about the extent to which an intervention is effective; since people display a low degree of numeracy, absolute outcome measures might hinder patients' understanding of effectiveness; thus, we ought to employ relative measures. The contrast between people's comparative understanding of relative versus absolute outcome measures is dubious. Relative measures, by promoting the base rate fallacy, fundamentally mislead patients into overestimating effectiveness.

The considerations here are concerned with the kinds of outcome measures that should be employed when summarizing data from clinical trials. The point of performing such experiments is to learn something about whether or not (and if so to what extent) an intervention will be effective for a broader target population and for particular patients. Once a trial has been performed and the data from the trial has been analyzed with an appropriate outcome measure, thereby determining an effect size for the intervention, the effect size is used to make an inference about the effectiveness of the medical intervention in a real clinical setting. This requires extrapolation.

8.4 Extrapolation

A widely held assumption is that the results of trials can be used to directly infer a general capacity of the intervention in question. Since the assumption is that the inferred capacity is general, one can infer that the medical intervention would manifest this capacity in a broader population and indeed in most particular patients. For instance, according to some in evidence-based medicine, in order to determine if one can extrapolate the results from trials to a particular patient, one should "ask whether there is some compelling reason why the results should not be applied to the patient. A compelling reason usually won't be found, and most often you can generalize the results to your patient with confidence" (Guyatt & Rennie, 2001). This is slightly more refined than simple extrapolation, since the guidance holds that one should determine if there are reasons why one should not extrapolate. Nevertheless, in the same breath the guidance claims that such reasons are rare, and thus the guidance amounts to simple extrapolation, most of the time, unless one is aware of a countervailing reason. I will call this methodological guidance 'simple extrapolation, unless' (SEU).[15]

Here is another expression of SEU, again from the evidence-based medicine community: "results of randomized trials apply to wide populations unless there is a compelling reason to believe the results would differ substantially as a function of particular characteristics of those patients." Similarly, an epidemiology textbook notes that "generalizing results obtained in one or more studies to different target or reference populations [is] the premier approach that public health professionals and policy makers use." One of the highest-profile guidance statements from methodologists in evidence-based medicine reiterates this view: "therapies (especially drugs) found to be beneficial in a narrow range of patients generally have broader application in actual practice."[16] The trouble with this claim is that, ironically, the 'evidence base' for it is extremely thin. Many of the articles that this group cites in support of this claim are merely opinion pieces in medical journals; the more rigorous empirical studies that they cite conclude that SEU is in fact problematic.[17] This defense of SEU is remarkable for its violation of its own evidence base.

There are a number of fundamental problems with SEU. First, it assumes that the default position should be that extrapolation is warranted, based on the further assumption that relevant differences between trial subjects and target patients are rare. The 'unless' clause in SEU states a condition, which, if satisfied, overrides the warrant for extrapolation. Post et al. (2013) note several ways in which such an overriding condition could be satisfied, including: if there are pathophysiologic differences in the illness under investigation that could lead to variability in treatment response, if there are differences in a particular patient compared with the experimental subjects that could diminish the treatment response, and if there are differences in patient or physician compliance that could diminish the treatment response. In the passage cited above, Guyatt and Rennie claim that the overriding condition is rarely satisfied—they assume that a particular patient is usually similar in all important respects to the subjects of a trial from which one wishes to extrapolate. However, one of their exception criteria that overrides warrant for extrapolation is the presence of differences between the target patient and the experimental subjects that may diminish the treatment response in the patient. In practice there are typically many such differences that may diminish the treatment response (Bartlett et al., 2005).

---

[15] Steel (2007) calls SEU 'simple induction': "Assume that the causal generalization true of the base population also holds approximately in related populations, unless there is some specific reason to think otherwise." My concern is specifically about extrapolation, so Steel's term is slightly misleading. Another bit of terminology: the term of art often used to describe those studies from which extrapolation is warranted is 'external validity.'

[16] The citations in this paragraph are: (Post, de Beer, & Guyatt, 2013), (Szklo & Nieto, 2007), and (Moher et al., 2010).

[17] One such article argues that trial design principles "limit the ability to generalize study findings to the patient population" (Gurwitz, Col, & Avorn, 1992), and another claims that "researchers, funding agencies, ethics committees, the pharmaceutical industry, medical journals, and governmental regulators alike all neglect external validity" (Rothwell, 2005).

Given the large number of criteria that trials employ that stipulate the properties that a subject must have (and other criteria that they cannot have) to be included in the trials, there are almost always differences between a particular real-world patient and the subjects in a clinical trial. Subjects in a trial are not drawn from a random sample of the broader population who have the disease in question, and the criteria that determine eligibility for a trial typically render subjects in a trial different in important respects from the broader population of people who have the disease. For instance, elderly patients, patients on other drugs, and patients with other diseases are excluded from trials. Some of these differences are known to modulate the effects of interventions. At the very least, the default assumption should not be that there are no such differences between trial subjects and target patients. For example, in the RECORD trial, which tested the safety of rosiglitazone, there were numerous inclusion and exclusion criteria applied to determine subject eligibility in the trial. The result was that the subjects in the trial were on average healthier than the target population.[18] The general phenomenon of employing selection criteria in a way that renders a subject population in a trial systematically different from a target population can be called 'recruitment bias.'

The second major problem with SEU is that it is unreliable due to forms of bias that transcend concerns about internal and external validity, such as publication bias. Even if there are in fact no substantial differences between the experimental subjects and target patients—and thus the overriding clause of SEU were not satisfied, and so extrapolation could be warranted—the results of published trials from which one is extrapolating could be entirely misleading, because the published trials may represent only a fraction of the trials that have been performed (Chapter 10). The reason that publication bias is a problem for SEU is that the subset of studies that are published will report a degree of effectiveness that is higher than the degree of effectiveness measured in all relevant studies (including those that are not published).[19] The subjects that are included in published studies differ from the set of all subjects in their responsiveness to the tested intervention, and under the safe assumption that the overall set of subjects (including subjects from unpublished studies) is more representative of target patients than the subset of subjects that are included in published studies with respect to their responsiveness to the intervention, publication bias threatens extrapolation. The presence of publication bias is a threat to any method of extrapolation that does not take the pernicious effects of publication bias into account. A method of extrapolation could take publication bias into account by decreasing estimates of effectiveness as measured in published studies when predicting the effectiveness of a medical intervention in a target population, and thereby improve on SEU (see Appendix 5).

---

[18] The subjects in the trial had a heart attack frequency of about 40 percent of the relevant group (middle-aged people with type-2 diabetes) in the broader population. See (Fuller, 2013a) for an articulation of this problem of non-representativeness of trial subjects. In Chapter 9 I argue that this problem contributes to the underestimation of the harm profile of drugs.

[19] This is a widely reported phenomenon. For an example, see (Eyding et al., 2010).

A problem with SEU that is closely related to the problem of publication bias is the fact that many results from trials are later overturned by contradictory results. Many findings that purport to show that an intervention is effective, or purport to show that an intervention is not harmful, are contradicted by results from subsequent research. SEU ignores the chance that the present evidence from which one extrapolates will be contradicted by later evidence.

The third major problem with SEU is that it ignores information regarding how the intervention works. Suppose a clinical trial reported that a particular medical intervention has an effect size of x, but that background knowledge of the mechanism of action for the intervention suggests that it would have a completely different effect incompatible with x. Further suppose that for a particular patient the overriding clause of SEU is not met—there are no reasons to suppose that the patient in question is different in any relevant respects from the experimental subjects of the trial. SEU tells us to infer that the intervention will cause x in this patient. This approach disregards the background knowledge of the mechanism of action of this medical intervention, and for that reason is misleading.[20] For example, in §8.2 I noted the example of high-dose chemotherapy, which appeared to be effective for breast cancer treatment in small trials (and which was known to be effective for treating blood cancers), but since chemotherapeutic drugs operate by interfering with mechanisms of cell division, cells that divide rapidly (like blood cells) are more susceptible to chemotherapy, and slower growing tumors (like breast tumors) are less susceptible to chemotherapy. The estimation of effectiveness of high-dose chemotherapy for breast cancer, based on positive results from the initial trials, should have been tempered by consideration of the intervention's mechanism of action.

The reliance on SEU contributes to overestimating the effectiveness of medical interventions, for reasons corresponding to the above problems with SEU. First, the features that real-world patients tend to have that render them different from subjects in trials—compared to subjects in trials, patients tend to be sicker, older, on more medications, and less compliant—usually mitigate treatment response in patients. Second, publication bias is asymmetric with respect to estimating effectiveness of medical interventions: trials that suggest a medical intervention is effective are far more likely to be published than trials that suggest a medical intervention is ineffective. Third, attention to the mechanism by which an intervention works ought, at least sometimes, to decrease one's estimate of the effectiveness of interventions.

As if SEU were not problematic enough, the 'unless' clause in SEU is often not attended to in policy development. Fuller (2013b) examined six clinical guidelines that recommend treatment with the five most commonly prescribed classes of medications

---

[20] Russo and Williamson (2007), Steel (2007), and Illari (2011), among others, argue that knowledge of the mechanism of action of a medical intervention is useful in warranting extrapolation about the intervention. This has been criticized by Howick (2011a) and Broadbent (2013) on the grounds that mechanistic knowledge is not necessary for extrapolation. All contributors to this debate, nonetheless, should agree that knowledge of mechanisms can sometimes aid in extrapolation.

for elderly patients in Ontario and found that these guidelines employ simple extrapolation for generalizing from trial results to treatment guidelines for wide target populations (including the elderly), without considering limits to generalizability.

The problems with SEU discussed here naturally suggest three corresponding correctives. First, extrapolation would be more reliable if subjects in trials were more similar to patients in the broader population. Not just any similarity will do, of course—it would not help if both the experimental subjects and all members of the target population were born under the sign of Scorpio (similarly, not just any difference between experimental population and target population invalidates an extrapolation). Second, when extrapolating from a particular measured value in published studies to make an inference of effectiveness in a target population, one should incorporate a subtraction factor to account for publication bias (and other relevant biases), the size of which should correspond to best estimates of the severity of bias in the relevant domain (see Appendix 5 for a generalization of this idea). Third, and related to the first, when extrapolating, one ought to consider knowledge of the mechanism of how the intervention works, and how the intervention can fail to work, and one ought to ensure that the target population is similar to the experimental population in all respects that are relevant to these mechanisms.

This latter principle has been characterized in various ways. To know that an intervention that appeared effective in an experimental setting will be effective in a target setting, Cartwright (2011) argues that we must know that the causal law that operated in the experimental setting also operates in the target setting, that the additional causal requirements that are necessary for the intervention to be effective (and that were present in the experimental setting) are in place in the target setting, and that the mechanism by which the intervention is effective operates in the target population and remains operative upon application of the intervention. Making a relatively similar point, Broadbent (2013) argues that we must know that we have eliminated potential interferers (where an interferer is a possible way in which an extrapolation could go wrong). A detailed proposal for a method of extrapolation is offered by Steel (2007), who argues that extrapolation can be grounded in 'comparative process tracing': identifying the relevant mechanism in the experimental population and comparing the mechanism in the target population, and assessing those stages in the mechanism that are most likely to be different between the two populations (although Steel's focus is on extrapolation from experiments on one species to knowledge about another species—typically humans—the method of comparative process tracing can also be valuable for extrapolation from research settings to target patient populations).

Crucially, when extrapolating measurements of effectiveness from a research setting to a clinical setting, one ought to take into account features of the research setting that go beyond mere concern about internal and external validity, including features such as the potential for publication bias and the chance that present findings will be contradicted by later research.

8.5 Measurement in Regulation

The three problems articulated above for measuring effectiveness are widespread in medical research. Evidential standards employed by regulatory agencies such as the U.S. Food and Drug Administration (FDA) are not suitable for addressing these problems. In this section I focus on the regulatory context in the United States. The Center for Drug Evaluation and Research (CDER) is a branch of the FDA that is responsible for regulating new drug approval. If a company wants to introduce a new pharmaceutical into the U.S. market, it must submit a 'new drug application' to CDER. The primary role of this institution is to evaluate the new drug application to determine if the new drug is (to use the FDA phrase) "safe and effective when used as directed." To determine if this is the case the FDA relies on measurements of effectiveness.

The institutions responsible for developing an experimental pharmaceutical (the 'sponsors,' who are usually pharmaceutical companies) must first test it on laboratory animals. If the results of animal tests are promising, the sponsors submit an 'investigational new drug application' to the FDA to get approval to begin human trials. Initial tests in humans are performed in 'phase 1' trials, which usually have fewer than one hundred healthy volunteers, and are used to discover the most harmful effects of the drug. If the drug appears to be not excessively toxic in a phase 1 trial, then the sponsor might move on to 'phase 2' trials. Phase 2 trials are RCTs, which usually involve a couple of hundred subjects, and are designed to test the efficacy of the pharmaceutical in patients with the disease meant to be treated. If the drug appears to have some efficacy in phase 2 trials, 'phase 3' trials are performed, which are also RCTs, but are larger than phase 2 trials, with several hundred to several thousand subjects, and are designed to gather more precise data on the efficacy of the experimental drug.

If a sponsor deems the drug promising based on phase 3 trials, they submit a 'new drug application' to the FDA. The FDA reviews the new drug application with a team consisting of physicians, statisticians, pharmacologists, and other scientists. If the new drug application is approved then the drug can be sold to consumers.

The epistemic standard for meeting the 'safe and effective' requirement is ultimately decided on a case-by-case basis. However, there are some common elements of the standard. The evidence must include a randomized trial in which the results are deemed 'positive.' The FDA definition of a positive trial is one in which a measured parameter in the experimental group of the trial is different than the parameter in the control group of the trial, the direction of this difference favors the experimental group, and this difference is deemed 'statistically significant' in that the p-value of a frequentist statistical test on this result is less than 0.05 (below I address the epistemological meaning of this). The FDA generally requires two positive trials to approve the new drug application, though it sometimes makes exceptions to the two-positive-trial rule, approving a new drug on the basis of a single trial, which might be supplemented with other confirming evidence, such as evidence from related positive trials or animal studies, and sometimes does away with required supplemental evidence if the trial happens to be a large multi-center trial. The measured parameter in an acceptable trial can be an important patient-level outcome (such as death), but trials that only measure 'surrogate endpoints' are also accepted (see Chapter 3 and §8.2). In short: to approve a new drug, generally the FDA requires two randomized trials in which the drug shows a statistically significant benefit.

There are numerous problems with this epistemic standard for drug approval, which render it too low. Although the epistemic requirements described above sound cumbersome, in the context of medical research they are too easy to satisfy with respect to any reasonable norm of evaluation—in other words, the FDA underregulates. That is the argument that follows.

A problem with the FDA standard is that because it is based on statistical significance it lends itself to 'p-hacking.' Spurious correlations can occur by chance, and the more complex a dataset is, and the more analyses performed on a dataset, the more likely one is to discover a spurious correlation. P-hacking can occur when a researcher exercises 'researcher degrees of freedom': researchers perform multiple studies, on multiple parameters, choosing which parameters to measure and which comparisons to make and which analyses to perform, and they can do this until they find a low enough p-value to satisfy the standard of statistical significance. Since low p-values are likely to occur by chance alone, p-hacking makes it easy to satisfy the standard of statistical significance even when the experimental drug is not in fact beneficial.[21]

Even when no p-hacking occurs, a statistically significant result in a trial does not entail that a clinically significant result has been found. This is for a number of reasons. The result, although statistically significant, may be due to chance. The result, although statistically significant, may be clinically meaningless because the effect size is tiny. The result, although statistically significant, may be clinically meaningless because the subjects in the trial differed in important ways from typical patients—as described above, trial subjects are very different from typical patients (typical patients tend to be older, on more drugs, and have more diseases than trial subjects, and these differences are known to modulate the effectiveness and harmfulness of pharmaceuticals).

Another problem with the FDA standard for drug approval is that although the effect size of a trial might be statistically significant, the measuring instrument in the trial might be biased (see §8.2). Recall my discussion of the HAMD scale above. The best assessments of antidepressants conclude that antidepressants on average lower HAMD scores by 1.8 points (Kirsch et al., 2008). But in any case this scale is a poor measuring instrument and its use contributes to overestimating the effectiveness of antidepressants—a drug's capacity to decrease one's HAMD score by 1.8 points does not indicate that the drug will be helpful in mitigating core symptoms of depression. The FDA standard is silent on this.

---

[21] P-hacking can be mitigated if trial designs explicitly state, in advance, what primary outcomes will be measured and how the data will be analyzed. Unfortunately, a recent study found that about half of clinical trials had at least one primary outcome that was changed, introduced, or omitted (Dwan et al., 2008).

Putting aside all of the above problems with the 'statistical significance' standard, there is a more technical and fundamental problem with this standard. To articulate this problem will require a brief use of formalisms. Suppose: our hypothesis of interest (H) is that a drug is effective, the null hypothesis (H_0) is that the drug is not effective, and a trial generates evidence (E) that suggests that the drug is effective with a p-value of 0.05. The FDA standard, which is satisfied in this case, is based on the probability that we would get E if H_0 were true: $P(E|H_0)$. But the FDA must determine if the drug is effective: the FDA must estimate how probable H is now that we have E: $P(H|E)$. There is a widespread habit of assuming that one can infer $P(H|E)$ from $P(E|H_0)$. But this is fallacious. To see this, apply Bayes' Theorem to $P(H|E)$:

$$P(H|E) = P(E|H)P(H) / \left[P(E|H)P(H) + P(E|H_0)P(H_0) \right]$$

(For a primer on Bayes' Theorem, see Appendix 1.) The statistical significance level only gives us the value of $P(E|H_0)$, which, as one can see by examining the equation above, is grossly insufficient to infer $P(H|E)$. Yet this is the epistemological basis of the FDA standard. Thus, the epistemological basis of the FDA standard is grossly insufficient for the inference it is required to make. Regulatory standards should be based on absolute measures of effectiveness (see §8.3), and of course should take into account statistical aspects of the measured properties, and should do so non-fallaciously.

Here is an illustration of another aspect of the poverty of the FDA's epistemic standard. In an FDA review of treatments for pediatric depression, fifteen trials that tested the effectiveness of seven leading antidepressants were analyzed. Twelve of these trials showed that the antidepressants were no better than placebo (Laughren, 2004). The report claimed:

We at FDA, however, do not view negative studies as proof of no benefit. In our view, absence of evidence for effectiveness in most of these programs does not constitute evidence of absence of benefit for these drugs...

This comment expresses the platitude often noted by statisticians that 'absence of evidence is not evidence of absence.' However, as Sober (2009) argues, this is often false: there are situations in which absence of evidence that H is true *is* evidence that H is false, namely, those situations in which were H true there would be at least a moderately high probability that you would know it were true.[22] Of course, if one was not doing anything to determine whether or not H were true—not gathering any evidence about H, say—then the statistician's platitude might apply. But clearly this is not the case for patented drugs: companies work very hard to gather evidence that suggests that their drugs are effective. There is a high probability that we would know H, if H was true. In Chapter 9 I argue that trials are tuned in many ways to sensitively detect possible benefits of pharmaceuticals (at the expense of being sensitive to the detection

---

[22] See also (Strevens, 2009).

of harms). When it comes to patented medical interventions, *absence of evidence of effectiveness is evidence of absence of effectiveness.*

Perhaps the most worrying problem about this regulatory standard is that it does not take into account publication bias. The two-positive-trials rule can be satisfied by a new drug application even if many trials generated evidence that suggested that the drug is not effective—as long as there are two positive trials, the standard is satisfied.[23] More generally, the FDA standard does not take into account the concerns regarding extrapolation raised in §8.4.

A survey of FDA drug reviewers indicated that even those involved in the drug review process believe that the epistemic standards for approval are too low—many drug reviewers expressed concerns about the low standards for evaluating effectiveness and harmfulness of drugs (Lurie & Wolfe, 1998). A reviewer claimed that the FDA leans toward approving everything. Reviewers reported cases in which they recommended that new drug applications be rejected and yet the drugs were nevertheless approved. In another context, a senior epidemiologist in the FDA claimed that the "FDA consistently overrated the benefits of the drugs it approved and rejected, downplayed, or ignored the safety problems... when FDA approves a drug, it usually has no evidence that the drug will provide a meaningful benefit to patients."[24] In Chapter 12 I return to the regulatory standard for drug approval using the concept of inductive risk, and suggest several ways in which such inductive risk could be better managed.

8.6 Conclusion

The measurement of effectiveness of medical interventions faces three challenges: the selection of a good measuring instrument, the use of an appropriate outcome measure, and the employment of a reliable method of extrapolating measurements in an experimental setting to a broader setting. The way these challenges are met in medical research is unsatisfactory, which systematically contributes to overestimating the effectiveness of medical interventions. These problems render current regulatory standards for drug approval too low.

---

[23] In a 1990 meeting this concern about the influence of publication bias on the FDA standard was articulated by a professor of statistics. The director of the FDA's Division of Neuropharmacological Drug Products at the time stated that by law the drug under discussion, sertraline (Zoloft), must be approved as long as there were two positive trials—regardless of the number of negative studies. See (Spielmans & Kirsch, 2014) for a critical review of FDA regulatory standards.

[24] (Graham, 2005). Thus far in the year of writing this chapter (September 2015), the FDA has rejected one drug and approved twenty-three, for an approval rate of 96 percent.

9

Hollow Hunt for Harms

9.1 Introduction

Harmful effects of medical interventions are systematically underestimated by medical research. This underestimation is a result of conceptual, methodological, and social factors present throughout the various stages of clinical research.

The difficulties with detecting harms of medical interventions begin with how harms are conceptualized and operationalized in clinical research (§9.2). After laboratory research, phase 1 trials are performed to evaluate the harm profile of experimental interventions. Unfortunately the vast majority of phase 1 trials remain unpublished, which generally and systematically skews the overall assessment of harm profiles of medical interventions (§9.3). If an intervention is deemed relatively harm-free in a phase 1 trial, it is tested in larger phase 2 and 3 trials. Randomized trials constitute one of the most significant hurdles in the hunt for harms. Most randomized trials are designed to be sensitive to the detection of potential benefits of medical interventions, and this sensitivity trades off against the sensitivity to detect potential harms of interventions. This is especially troublesome given that trials are usually thought to produce the best evidence regarding the effects of medical interventions (though in Chapter 5 I note criticisms of this view). In §9.4 I highlight several ways that trials are designed such that the harms of interventions are underestimated. Once a medical intervention has been approved for use in the clinical setting, harms are hunted with the use of passive surveillance and sometimes with phase 4 trials. This has both practical and epistemic shortcomings (§9.5).

The hunt for harms is embedded in a social nexus that exacerbates the underestimation of harms. Most evidence regarding harms of medical interventions is generated by trials that are funded and controlled by the manufacturers of the interventions under investigation, and whose interests are best served by underestimating the harm profile of interventions. This leads to widespread limitation of the evidence regarding harms that is made available to independent scientists, policy-makers, and the public, and this in turn contributes to the underestimation of the harm profiles of medical interventions. Regulators lack the authority that would allow them to properly estimate harm profiles of interventions, and frequently contribute to shrouding the relevant evidence regarding harms in secrecy (§9.6). In Chapter 11

(§11.2) I list many examples of drugs that have been withdrawn from the market due to excessive harms.

The net effect of these conceptual, methodological, and social factors is that our available medical interventions appear to be safer than they truly are. Were these factors mitigated in medical research, the harm profiles of interventions would be more truthfully represented, and would be deemed more harmful than they now are.

9.2 Operationalizing Harm

A harm of a medical intervention is, of course, an effect of the intervention, just as a benefit of an intervention is an effect. The interpretation of an effect *as* a harm (or conversely, as a benefit) is a normative judgment, and as such is influenced by social values (see Chapter 2). Such judgments are not always straightforward. A compelling illustration is provided by the drug methylphenidate (Ritalin), often prescribed to treat attention deficit hyperactivity disorder (ADHD). The alleged benefits of methylphenidate depend on a particular social nexus and are conceptually intertwined with its harms. Empirical tests of methylphenidate suggest that it mitigates children's bodily motions and frequency of social interactions, which might be seen as a benefit by an overworked teacher. But this effect could be judged as harmful by someone who thinks that children moving around, playing, and socializing are positive behaviors. As one critic puts it, stimulants like methylphenidate "work for the teacher," but not necessarily the child.[1] The same effect of an intervention may be considered a benefit or a harm depending on one's broader normative commitments and sociocultural context. There are, though, many effects of medical interventions that can be considered harms with little ambiguity in typical cases, from insubstantial effects such as a minor headache, to more severe effects such as death.

Harms are often thought of as discrete outcomes, referred to as adverse events, or if they are extremely harmful, serious adverse events. A harm, however, can be a small change of a continuous parameter rather than a change of a discrete parameter; many harms should not be thought of as events, since discrete events constitute only a fraction of the potential harms of a medical intervention. As I discuss in §9.5, the vast majority of harm data comes from passive surveillance and observational studies, in which a particular token event can only be detected as a harm in the first place if a patient or a physician observes and interprets an effect of an intervention as a harm, and reports it as such. Small effects and common effects are often not reported. If a drug causes a patient to gain several pounds, this effect could easily go unnoticed by the patient and the physician, and even if it were noticed, there is no way that the

---

[1] (Whitaker, 2010). Unfortunately, our best evidence suggests that methylphenidate does not work for the child. Self-reports of well-being and assessments of academic performance are not improved by methylphenidate, and in the longer term, methylphenidate causes worse outcomes (see §9.4).

patient or the physician could reliably assess the drug as a cause of the weight gain. In other words, such an effect probably would not be attributed as an effect of the drug, and if it were, such an attribution would not be reliable.[2]

Terminological choices contribute to obscuring the harm profiles of medical interventions. Concerns about harms of interventions are often referred to with terms such as 'drug safety.' A report of a new kind of harm of a medical intervention is a 'signal' of a 'safety finding,' which is documented via 'safety reporting.' For example, the FDA, when talking about serious harms such as death and strokes caused by peroxisome proliferator activated receptor (PPAR) modulators (described below) referred to these events as 'clinical safety signals,' and some drugs in this class were removed from the market because of 'clinical safety.' The use of the term *safety* to refer to *harm* is perhaps the most egregious Orwellian locution in medicine. Moreover, other benign-sounding phrases employed in discourse about medical interventions, such as 'side effects' or 'adverse events'—which can refer to collapsing lungs, self-mutilation, exploding tendons, and death—contribute to the opacity of harms of medical interventions.

The way harms are operationalized in clinical research contributes to their underestimation. For example, before it was established that antidepressants can cause suicidal ideation, some analyses of trial data suggested that these drugs do not in fact cause this terrible harm. The data from these trials were of patient outcomes measured with the Hamilton Rating Scale for Depression (HAMD). In Chapter 8 I argue that this is a poor instrument for measuring the effectiveness of antidepressants. But the HAMD is an even worse instrument for measuring certain harms of antidepressants, including suicidality. There is a single question on this scale regarding suicidality, as follows:

0 = Absent.
1 = Feels life is not worth living.
2 = Wishes he were dead or any thoughts of possible death to self.
3 = Suicidal ideas or gesture.
4 = Attempts at suicide (any serious attempt rates 4).

The problem is that an antidepressant could cause a patient who has occasional suicidal ideas to develop severe and frequent suicidal ideation and self-mutilation without actually attempting suicide, and the patient's score on this question would not change (since both before and after the antidepressant the patient would receive a score of 3 on suicidality). The HAMD is insensitive to such harms. Since this tool is the primary measuring instrument employed in trials of antidepressants, trials systematically underestimate this harm of antidepressants.[3] This illustrates the more general

---

[2] This also holds, in principle, for small effects that are beneficial. However, as I argue in §9.4, clinical trials are typically designed to be more sensitive to detecting benefits than they are to detecting harms.

[3] An egregious example of the use of this scale to test for increase in suicidality caused by antidepressants is in (Beasley et al., 1991), in which 'emergence of substantial suicidal ideation' was operationalized as a change in the score on the HAMD suicidality question from 0 or 1 at the start of a trial to 3 or 4 at the end of the trial. Unsurprisingly, this study was funded by Eli Lilly, the manufacturer of fluoxetine (Prozac), and the lead author was an employee of the company.

point that the way harms are operationalized in clinical research can contribute to the underestimation of the harm profiles of medical interventions.

Another example of how the operationalization of harms contributes to their underestimation is about the drug rosiglitazone (Avandia), once the world's best-selling drug for type 2 diabetes. By 2007, evidence was mounting that rosiglitazone increases the risk of cardiovascular disease and death. The manufacturer of rosiglitazone funded a large trial in an attempt to disprove this.[4] The primary outcome measured in this trial was a composite outcome that included all hospitalizations and deaths from any cardiovascular causes. This outcome included cardiovascular hospitalizations that were very likely not related to the interventions (rosiglitazone or control), and thus, because we can presume that hospitalizations and deaths that were not caused by either intervention occurred at roughly the same rate between the trial groups, and that hospitalization is a much more frequent event than death, the overall larger number of this outcome in both groups minimized the relative difference in outcomes observed between the groups. That is, the outcome of interest—cardiovascular disease and death caused by rosiglitazone—was, in both groups, 'watered down' by including the much more frequent outcome of hospitalization, making it less likely to detect a statistically significant difference between the groups.[5] The way that the potential harm was operationalized in this trial artificially lowered the chance of detecting harms of the drug.

The broader point illustrated by the examples above is that harms of medical interventions will only be found if they are properly looked for. Operationalizing a harm in certain ways—such as by employing a measuring instrument or an outcome that is insensitive to the harm in question—amounts to not properly looking for the harm.

9.3 First-in-Human, Never Seen Again

A phase 1 first-in-human study is an experiment in which an intervention is administered in humans for the first time. Generally, medical interventions first are evaluated with in vitro and animal experiments, and if such experiments provide evidence to think that the intervention might be safe and effective for human use, a first-in-human study is performed. Such studies are referred to as phase 1 trials.

Such trials are risky for the subjects. A recent phase 1 trial of an experimental molecule tested a dose of the molecule that was 500 times lower than the dose found to

---

[4] This was the RECORD trial.

[5] Moreover, the outcome 'hospitalization' obviously depends on a patient being hospitalized, but this is a socioeconomic decision as much as a health outcome, and the trial included patients in dozens of countries with diverse hospitalization practices. This diversity could have introduced variability in the data, which would make it less likely that a statistically significant difference between groups would be detected, even if there was a difference.

be safe in animals.[6] The six men who were given the drug quickly developed intense headaches, back pain, intestinal pain, diarrhea, fevers, low blood pressure, and lung pain, and after forty-eight hours each had multiple organ failures.

Despite the risk of phase 1 trials, they are important because they provide the foundation for assessing harms of medical interventions. Given that phase 1 trials are the first time an experimental intervention is tested in humans, they provide crucial evidence regarding harms. Such evidence is relevant, obviously, to the harm profile of the particular molecule under investigation, but such evidence is also relevant to the harm profile of the class of molecules to which the particular molecule belongs, and is more broadly relevant to the harm profile of drugs, generally. I will use the following notation in the arguments to come: molecule x is a member of the class of molecules of type T, and this class is itself a member of the class of all drugs D. Evidence from a phase 1 trial on x is relevant, obviously, to the harm profile of x, but is also relevant to the harm profile of T (albeit more indirectly), and is also relevant to the harm profile of D (more indirectly still). Evidence from phase 1 trials is, therefore, hugely important for our knowledge of the general harmfulness of drugs.

Unfortunately, such evidence is rarely shared publicly. The vast majority of phase 1 trials are not published.[7] It is difficult to know exactly what proportion of phase 1 trials are published because there is no registry of what molecules have been tested by such trials. Empirical studies of the publication bias of phase 1 trials have relied on records of institutional review boards (committees of universities and hospitals for reviewing proposed experiments involving humans), and it appears that about 95 percent of phase 1 trials are not published. Cases like the one above, in which the public learns about the harm profile of a molecule because of a tragedy in a phase 1 trial, are unusual. For experimental interventions with less dangerous harm profiles we know very little about their harm profiles, because the evidence regarding the harm profile of the vast majority of molecules is rarely published or publicized.

Those molecules that are not obviously harmful in a phase 1 trial typically go on to be tested in phase 2 and phase 3 trials, and thus the broader scientific community can infer that such molecules have passed a phase 1 trial, and so might infer that such a molecule is at least somewhat safe (though, as I argue below, that inference is unwarranted). Those molecules that appear to be relatively harmful in phase 1 trials rarely go on to be tested in further trials, and such phase 1 trials are rarely published. This publication bias of phase 1 trials is wasteful. Future scientists who are unaware of the harm profile of x or other molecules of class T, for which prior phase 1 trials have been performed, and who want to know the harm profile of x or another member of T, are liable to perform wasteful subsequent phase 1 trials. This also has the potential for

---

[6] This drug was called CD28-SuperMAB (also referred to as TGN1412). See Lemoine (2017) for an insightful analysis of this case. The dosage given to the human subjects relative to the safe dosage in animals was in fact an estimate based on extrapolation from an animal model of CD28-SuperMAB, since this class of drug is species-specific.

[7] For empirical evidence on this, see (Decullier, Chan, & Chapuis, 2009).

causing needless harm to subjects in these subsequent phase 1 trials. There is, though, a consequence of publication bias of phase 1 trials that runs much deeper than this.

When assessing the harm profile of a molecule (x), if one is unaware of past evidence regarding harms (of x and more generally T), then one's prior probability that the molecule is harmful will be lower than it should be (that is, lower than it otherwise would be if one was aware of such evidence). Since molecules that appear safe in phase 1 trials tend to be evaluated in larger and more public phase 2 and 3 studies, and molecules that appear harmful do not, and since most phase 1 trials are not published, it follows that, of all drugs that are tested for clinical use, the proportion that *appears* harmful is lower, perhaps much lower, than is truly the case.

This is crucially important, so I reiterate the argument in more formal terms. Our assessment of the harm profile of x can be represented as a conditional probability, $P(K|E)$, where K is the hypothesis that x is harmful, and E is relevant new evidence regarding the harm profile of x (E could be data from a phase 1 trial, or from a phase 3 trial, or whatever). Our assessment of the harm profile of a molecule is directly proportional to $P(K)$, the prior probability that the molecule is harmful (see Appendix 1). How ought one determine $P(K)$? This is a notoriously difficult question for scientific methodology. But in this context there is at least an obvious constraint on an answer.

The prior probability that x is harmful depends on past evidence regarding x and other molecules like it, including molecules of type T and more broadly all drugs D. We have access to only a small subset of the relevant past evidence regarding harms of x and T and D. Given the rampant publication bias noted above, the evidence that we do not have access to is more likely to confirm K than the evidence that we do have access to. It follows that our assessment of $P(K)$ is artificially low, due to publication bias, and would be higher and more accurate if we had access to all relevant evidence. This, concomitantly, would have a direct positive impact on $P(K|E)$. This is a general argument. Thus, *for all drugs*, our estimate of the probability that any particular drug is harmful is artificially lower than it otherwise would be if we had all the relevant evidence from phase 1 trials. The extent of this problem is difficult to estimate, but given the empirical estimates of the frequency of publication bias of phase 1 trials, it appears to be extremely serious.

What is the appropriate reference class for assessing the harm profile of x? Should we compare x only to other molecules like it, or to all molecules of type T, or all drugs D? Like the question about assessing $P(K)$, this is a notoriously difficult question for scientific methodology. Again, though, in this context there is a straightforward constraint on an answer. Since x has some close similarities with other molecules of type T, and broad similarities with other members of D, when assessing the harm profile of x at the very least one should take into account the harm profiles of more members of T and D than is currently possible due to publication bias. And since, as above, the publication bias regarding the harm profiles of T and D is systematically skewed toward underestimating harms, it follows that if one were able to assess the harm profile of x

with a more appropriate reference class, the harm profile of *x* would be more accurately assessed and *x* would appear more harmful than it otherwise does.

Recall rosiglitazone, which is a modulator of proteins called peroxisome proliferator-activated receptors (PPARs) that regulate the expression of genes. In recent years, more than fifty PPAR modulators have failed clinical tests, and many of these failures have been due to harms caused by the PPAR modulators. Evidence of such harms was available even prior to phase 1 trials: for example, PPAR modulators were found to cause numerous types of tumors and cardiac toxicity in rodents. Our estimations of the harm profile of an experimental intervention should take such background knowledge into account.[8]

Another factor that ought to influence one's assessment of the prior probability that *x* is harmful is background knowledge of the way that *x* intervenes in normal and pathological mechanisms. PPAR modulators are again a good example: any given PPAR modulator can influence the expression of many dozens of genes, and thus we should expect a diverse range of effects from using such a drug.[9] Unfortunately, this kind of knowledge is often downplayed (see Chapter 5)—mechanistic reasoning is typically denigrated by evidence-based medicine. But recall my discussion of the cascading complexity of effects of pharmaceuticals in Chapter 4: we should expect numerous harmful effects of many drugs because of their complex causal effects in our normal physiology. Taking account of knowledge of this cascading complexity of effects of drugs ought to increase our estimation of the harm profiles of drugs.

Given the argument presented in this section, one would expect to see examples of drugs that appear to be relatively safe based on evidence from phase 1 trials, but then come to be viewed as relatively harmful based on evidence from clinical trials and post-market surveillance (phases 2–4). And this is precisely what we observe. Just among the class of PPAR modulators there are many such examples: troglitazone has been withdrawn in some jurisdictions because it appears to cause liver damage, tesaglitazar has been withdrawn in some jurisdictions because it appears to cause elevated serum creatinine, pioglitazone has been withdrawn in some jurisdictions because it appears to cause bladder cancer, and muraglitazar has been withdrawn in some jurisdictions because it appears to cause heart attacks, strokes, and death.[10]

Publication bias of other study types further contributes to the systematic underestimation of the harms of medical interventions, as I discuss in §9.6. In short, the lack of availability of evidence from phase 1 trials contributes to the systematic underestimation of harms of medical interventions.

---

[8] Unfortunately, according to a leading type-2 diabetes researcher, "few publications have detailed the precise toxicity encountered" (Nissen, 2010), and "few data on toxicity are available in the public domain because of the common industry practice of not publishing safety findings for failed products" (Nissen & Wolski, 2007).

[9] In the words of (Nissen, 2010): "the effects of these agents are unpredictable and can result in unusual toxicities."

[10] By 'withdrawn' here I mean that the particular drug has been removed from a national jurisdiction based on the noted harm. Some of these drugs are still available in some jurisdictions.

9.4 Clinical Trials and the Abuse of Power

Trials are not good methods for hunting harms of medical interventions.[11] In principle randomized trials could be reliably employed to hunt for harms, though such trials would have to be larger and longer than most trials performed today and incorporate other more fine-grained methodological changes. In practice randomized trials are designed to be sensitive to the detection of benefits of medical interventions at the expense of being sensitive to the detection of harms.

To make this argument I employ the concept of statistical power. The statistical power of a trial is characterized in several ways: the probability that a statistical analysis of data from a trial rejects a false null hypothesis; the probability of avoiding a 'type II' error (falsely concluding that there is no difference between the experimental group and the control group of a trial); the probability of detecting a difference between the experimental group and the control group of a trial if there is truly a difference to be detected. Broadly construed, power refers to the sensitivity of a trial to detect an effect of the intervention under investigation, when there is such an effect to be detected. The statistical power of a trial depends on three parameters: the effect size of the intervention under investigation, the number of subjects in the trial, and the variability of the data. It is usually difficult to achieve satisfactory power in trials of medical interventions, for a variety of reasons: many medical interventions have small effects, increasing the number of subjects in a trial is expensive, and subjects can respond in very different ways to experimental interventions.

Trial designers try to maximize power in a number of ways. One is to maximize the observed effect size in a trial by including only subjects who are most likely to show the most benefit of the intervention. The parameter that influences power that is often easiest for trial designers to control is the variability of the data: to minimize variability, trial designers include a relatively homogeneous group of subjects. The greater the similarity among subjects with respect to parameters that can influence outcomes of a trial (such as age, sex, or the presence of other diseases), the less variable will be the data. Finally, despite the expense, trials will often include many thousands of subjects. (There is, obviously, great financial incentive to avoid the error of falsely concluding that a potential new medical intervention is ineffective.)

To maximize the observed effect size and minimize the variability of data, trial designers employ various criteria constraining what subjects are included or excluded from the trial. For example, it is common to exclude elderly subjects, subjects on other drugs, or subjects with other diseases (see Chapter 8). The most egregious of these trial design features are called enrichment strategies: after the enrollment of subjects, but prior to the start of data collection, subjects are tested for how they respond to placebo

---

[11] As Michael Rawlins puts it in his celebrated *Harveian Oration*: "in the assessment of harms RCTs are weak at providing evidence…they are an unreliable approach to the definitive identification of harms" (2008).

or the experimental intervention, and those subjects that do well on placebo or those subjects that do poorly on the experimental intervention are excluded from the trial.[12]

One result of these strategies is that subjects of trials are different in many important respects from the patients who use interventions once they are approved for use in a clinical setting. Some of these differences are known to influence the harm profiles of medical interventions. Older people, pregnant women, and patients on other drugs (for example) are more likely to be harmed by medical interventions, but they are also precisely the kinds of people who are excluded from trials. For example, the most common harm of statins is myopathy, which ranges from simple muscle pain and weakness, to rhabdomyolysis, which is a severe condition in which muscle tissue dies and releases proteins (myoglobin) into the blood, which can cause kidney failure and death (other harms of statins include stroke, congenital defects, diabetes, cancer, neuromuscular symptoms, nerve damage, abnormal liver function, joint problems, and tendon damage). This risk is higher among women, elderly people, and people with other conditions like infections, seizures, and kidney disease—precisely the kinds of people that are excluded from clinical trials. One aim of excluding such patients is to maximize the ability to detect potential benefits of a drug, but the exclusion of such patients also minimizes the ability to detect potential harms of a drug. This is recruitment bias, noted in Chapter 8.

Here is an example. Worrall (2010) notes that in the large ASSENT-2 trial, an exclusion criterion was "any other disorder that the investigator judged would place the patient at increased risk." Of course, there is an ethical basis for this, namely, the protection of subjects who are more likely to be harmed by experimental interventions. However, this exclusion criterion directly mitigated the ability of the trial to detect the harms of the intervention that would result when the intervention is employed in a real-world clinical setting, since it is precisely in the clinical setting in which patients have other disorders that put them at increased risk of harm. The enrichment strategy that involves excluding subjects who fare poorly on the test drug is another trial design feature that mitigates the ability of a trial to detect harms of medical interventions, because those subjects who in fact experience harms caused by the intervention are excluded from the trial.

Usually, trial power refers to the ability of a trial to detect a benefit. However, since a harm of a medical intervention is simply another effect of the intervention, just like a benefit, the power of a trial can also refer to the ability of a trial to detect harms. I will call the sensitivity of a trial to detect benefit power$_B$ and the sensitivity of a trial to

---

[12] One enrichment strategy is called a run-in period, which involves the exclusion of placebo-responders before the trial begins. Here is an example mentioned in Chapter 8. Of fifteen trials analyzed by the FDA regarding antidepressant use in children, only three showed positive results. Two of these three studies were of fluoxetine (Prozac), and thus fluoxetine was approved for use in children diagnosed with depression. However, the trials put all children on a placebo for one week, and any children who significantly improved during this week were excluded from the trial. This rendered the subjects in the trial different from real-world patients.

detect harm power$_H$. Power$_B$ and power$_H$ trade off against each other. There are numerous ways in which trial designers maximize power$_B$ at the expense of power$_H$. The exclusion of certain kinds of patients and inclusion of other kinds of patients, mentioned above, is one such strategy. The net result is that the power of trials (and more broadly the sensitivity of trials) to detect harms is typically much lower than the power of trials to detect benefits.[13]

To return to my running example, a meta-analysis was performed that showed that rosiglitazone causes an increased risk of heart attack and death from cardiovascular disease (Nissen & Wolski, 2007). The individual trials that this group amalgamated were too small to have an adequate power to detect this rare but severe harm. The RECORD trial was an attempt to show that rosiglitazone does not increase the risk of heart attacks (I noted this trial in Chapter 8 in my discussion of extrapolation). This trial employed seven inclusion criteria and sixteen exclusion criteria, and 99 percent of the subjects were Caucasian. One result of these criteria was that subjects in the trial were, on average, healthier than the broader target population; for example, subjects in the trial (that is, in both the control group and the rosiglitazone group) had a heart attack rate about 40 percent less than the heart attack rate in the equivalent demographic group (middle-aged people with type 2 diabetes) in the broader target population.

If subjects in the experimental group of a trial withdraw from the trial due to harms of the experimental intervention, then the presentation of the trial data could give a misleading impression that the intervention is safer than it actually is, because data about harms in those subjects who withdrew is not collected. Unfortunately there is scant evidence about the frequency of subject withdrawals, and insufficient reporting of subject withdrawals is ubiquitous.[14]

There are two other limitations of trials that contribute to the underestimation of harms of interventions: their size and their duration. Trials normally enroll enough subjects to detect the potential benefit of the intervention. Any more subjects adds expense. However, this number of subjects is often not enough to detect harms that are severe but rare. Trial size is optimized to achieve satisfactory power$_B$, without concern for power$_H$. The duration of a trial is normally just long enough to detect the potential benefit of the medical intervention. Some studies of antidepressants, for example, only evaluate the drugs for a period of weeks. A longer trial adds expense. However, some harms of drugs manifest only after years of taking the drug. Methylphenidate, for

---

[13] Tsang, Colley, and Lynd (2009) performed calculations of the power to detect serious harms of interventions on publications that did not originally report the power to detect harms (though the publications did report that no statistically significant serious harms were found). When they calculated power$_H$ for the trials, they found values ranging from 0.07 to 0.37. Thus, the probability is very high that these trials would falsely report that there are no harms of the interventions even if there were in fact harms.

[14] For example, in a review of 133 publications of RCTs published in 2006 in six leading general medical journals, Pitrou, Boutron, Ahmad, and Ravaud (2009) found that no information on severe adverse events was given in 27 percent of the articles, and no information on subject withdrawal due to adverse events was given in 47 percent of the articles.

example, has been shown to cause stunted growth in children, but this is only found three years after the initiation of treatment with the drug.[15] For these two reasons—the small size and short duration of trials—larger and longer observational studies are usually relied on to detect harms, but as I note below, observational studies have their own practical and epistemic shortcomings.

Since trials underestimate harms, we should expect to observe examples of drugs that appeared to be relatively safe after clinical trials but came later to appear to be more harmful once used in a clinical setting. This phenomenon is widespread. The worst cases are those in which medical interventions are pulled from the market by manufacturers or regulators. Here are a few examples from the last several years: valdecoxib (Bextra), fenfluramine (Pondimin), gatifloxacin (Gatiflo), and rofecoxib (Vioxx). Other cases are those in which the harm profile in the clinical setting appears worse than randomized trials suggested, but have been left on the market for whatever reason (often, regulators consider the benefit-harm profile of the drug to remain favorable regardless of the increasing estimation of its harm profile). A few examples include: celecoxib (Celebrex), alendronic acid (Fosamax), risperidone (Risperdal), olanzapine (Zyprexa), and in the United States, the running example of this chapter, rosiglitazone (Avandia) (see §11.2 for more examples).[16] Some of these drugs have been the subject of massive lawsuits because the manufacturers deceptively downplayed the harm profiles of the drugs. Another important property of trials that contributes to the underestimation of harms of medical interventions is not intrinsic to the methodology of the trials themselves, but is rather about how the evidence from the trials is shared publicly and used by regulators (see §9.6).

In short, clinical trials are not usually well designed to hunt for harms.[17] In this section I have argued that two important methodological properties of trials—$power_B$ and $power_H$—trade off against each other, usually in favor of $power_B$ at the expense of $power_H$, and this trade-off is constituted by a plurality of fine-grained methodological choices made by trial designers.

---

[15] Methylphenidate stunts growth by as much as 2 cm in height and 2.7 kg in weight (Swanson et al., 2007). Here is another example. For decades cortisone shots have been used to treat tendinopathy (overuse injuries like tennis elbow). Most of the trials testing this intervention were short, and cortisone appears to offer short-term pain relief. But a recent meta-analysis found that in the longer run patients receiving cortisone injections had a lower rate of recovery than those who did nothing or got physical therapy (Coombes, Bisset, & Vicenzino, 2010).

[16] The example of rosiglitazone demonstrates the importance of meta-analysis in the hunt for harms: the individual trials were too small to demonstrate a harmful effect of rosiglitazone, but the amalgamated evidence was able to demonstrate the harm. Though meta-analysis has its own shortcomings, as argued in Chapter 6, the best meta-analyses are those that are able to get access to both published and unpublished data, as Nissen and Wolski (2007) were able to.

[17] A survey of 142 randomly selected reports of clinical trials of psychiatric interventions found that only a fraction bothered to address harms, and on average, reports of trials used 1/10 of a page in the results section to discuss harms (Papanikolaou, Churchill, Wahlbeck, & Ioannidis, 2004).

9.5 Jump Now, Look Later (but Don't Look Hard)

The vast majority of data regarding harms of medical interventions comes from observational studies and passive surveillance conducted after a given medical intervention
has been approved for clinical use. These studies are called phase 4 trials and post-
market studies. The fact that the majority of data regarding harms of medical interventions comes from post-market studies has an important practical consequence, and
the fact that such studies are usually observational designs has an important epistemic
consequence.

The bar that a new medical intervention has to get over in order to be approved for
consumption and marketing is low. The FDA, for example, requires only two randomized trials that show that a new medical intervention has some beneficial effect,
regardless of how many trials were performed on the intervention, and despite the fact
that the power$_H$ of such trials is usually extremely low (see Chapters 8 and 12 for criticism of this standard). So, at the point at which a new intervention is approved for
general use, there is scant evidence available on the harm profile of that intervention.
After the intervention is approved for clinical use, its harms are assessed by passive
surveillance systems and observational studies. Such interventions are consumed by
typical patients, often numbering in the millions. It is only at this point, when new
interventions are used in clinical settings, rather than an experimental setting, that
most data on harms is gathered. This data comes from patients who inadvertently
become subjects in a study regarding the harm profile of the drug. Without knowing it,
such patients are unwitting guinea pigs in the hunt for harms.

There is reason to think that post-market surveillance severely underestimates
harms of medical intervention. One empirical evaluation of this puts the underestimation rate at 94 percent (this was based on a wide-ranging empirical survey by (Hazell &
Shakir, 2006)).

Unfortunately, because observational studies and passive surveillance do not
involve a randomized design, they are typically denigrated relative to randomized
trials. Since most evidence regarding harms of medical interventions comes from non-
randomized studies (especially rare severe harms), the dominant view of EBM thereby
denigrates the majority of evidence regarding harms of medical interventions (see
Chapter 5 for criticisms of this).

This view has influenced regulators. For instance, this passages comes from testimony of a senior epidemiologist at the FDA, during a congressional hearing regarding
the drug rofecoxib (Vioxx):

The corporate culture within CDER is also a barrier to effectively protecting the American
people. The culture is dominated by a world-view that believes only randomized clinical trials
provide useful and actionable information and that postmarketing safety is an afterthought.[18]

---

[18] CDER, the Center for Drug Evaluation and Research, is part of the FDA. The full testimony, which is
a scathing account of the regulation provided by the FDA, is available online.

According to the line of thinking criticized in this testimony, only randomized trials can provide compelling evidence regarding harms of interventions, and since the majority of data regarding harms comes from non-randomized studies, and since the data regarding harms that does come from randomized trials is fundamentally limited for the reasons I described above, U.S. regulators, by their own lights, have a paucity of reliable evidence regarding harms of interventions.

An argument has been raised by Vandenbroucke (2008) against the view that, in the hunt for harms, randomized trials are better than observational studies. Because harms of drugs are unintended and often have unknown effects, physicians cannot bias treatment allocation with respect to such effects (this is selection bias—see Chapter 10). Thus, selection bias is less of a worry for unintended harmful effects as it is for intended beneficial effects, and so one of the central advantages of randomized trials over observational studies is mitigated in the hunt for harms (see also (Osimani, 2014)). The upshot is that observational studies do not typically *overestimate* harm profiles of drugs. Indeed, there is some empirical evidence suggesting that observational studies *underestimate* harm profiles.

One group compared estimates of harms from large randomized trials that included a thorough hunt for harms to equivalent non-randomized studies, and found that non-randomized studies, on average, have conservative estimates of harms of interventions relative to randomized trials on the same intervention.[19] One reason for such a finding is that those patients who take their medications on a schedule that is faithful to their physicians' orders tend to be healthier than non-compliant patients, and thus there is a confounding factor when comparing the outcomes of patients who consume more of an intervention compared with patients who consume less of it. Those who consume medications more faithfully happen to be healthier than those who consume medications less faithfully. Those who are healthier anyway experience better outcomes and fewer harms when on the medication. Thus, observational studies that compare patients who consume more medical interventions than other patients tend to overestimate the benefits of interventions and underestimate their harms. For the above reasons, even if regulators do not typically have access to evidence regarding harms of medical interventions from randomized trials, they could rely on evidence from non-randomized studies and be confident that, on average at least, they are not overestimating the harm profile of medical interventions.

The evidence regarding harms that regulators do have access to, however, is apparently good enough to keep secret, a subject I turn to in the next section.

9.6 Secrecy of Data

A vast amount of evidence regarding harms of medical interventions is shrouded in secrecy. Companies that pay for trials claim that they own the data from the trials, and

---

[19] (Papanikolaou, Christidi, & Ioannidis, 2006).

researchers who participate in industry-sponsored trials are often bound by gagging clauses in their contracts that constrain their ability to share data, even if they suspect that an intervention causes harm.

Consider reboxetine, which is an antidepressant sold in Europe during the past decade. Recently a meta-analysis was performed in which the researchers had access to both published and unpublished data.[20] Of the thirteen trials that had been performed on reboxetine, data from 74 percent of patients remained unpublished. Seven of the trials compared the drug against placebo: one had positive results and only this one was published; the other six trials (with almost ten times as many patients as the positive trial) gave null results, and none of these were published. The trials that compared reboxetine to competitor drugs were worse. Three small trials suggested that reboxetine was superior to its competitors. But the other trials, with three times as many patients, showed that reboxetine was worse than its competitors on the primary outcome, and had worse side effects. Just like phase 1 trials, phase 3 trials suffer from rampant publication bias, which results in the exaggeration of benefits of medical interventions (Chapter 8) and the underestimation of harms.

The tribulations of rofecoxib (Vioxx) provide a striking example of such secrecy, which was later publicly exposed. The manufacturer of rofecoxib carried out the VIGOR trial to test the drug's safety and efficacy. It is now widely thought that in the landmark publication of this trial the authors withheld data on cardiovascular harms of rofecoxib. This was the view of the editors of the journal that published the article (*The New England Journal of Medicine*) after they learned of company memos that showed that at least two of the article's authors were aware of the data on cardiovascular harms. The methodological issue was portrayed by the company (Merck) and the article's authors as more subtle: the analysis of cardiovascular harms followed a predefined plan, according to which the study stopped collecting cardiovascular harm data on a particular date, and so they claimed that it would have been ad hoc and thus inappropriate to include the reports of cardiovascular harms associated with rofecoxib that were gathered in the two weeks after this cutoff date.[21] This is controversial: many philosophers of science hold that the timing of when one gains evidence is irrelevant to how confirmatory that evidence is. Regardless, at the very least the particular evidence regarding cardiovascular harms that was in fact gathered after the cutoff date could have been made public. Instead, such data were kept secret for too long.

Examples of the secrecy surrounding evidence of harms of medical interventions are easy to find. Here are three other examples of very widely used drugs. Olanzapine

---

[20] (Eyding et al., 2010). For more details of this case, see (Goldacre, 2012).

[21] The relevant citations are (Bombardier et al., 2000), (Curfman, Morrissey, & Drazen, 2005), and (Bombardier et al., 2006). The cutoff date for cardiovascular harms was February 10, 2000, while the cutoff date for reporting data on gastrointestinal events—the beneficial parameter in question—was March 9, 2000, and so the trial had more time to gather data on benefits than it did on harms. Thus this trial provides another example of the thesis presented in §9.4. See (Biddle, 2007) for a detailed account of what he calls the "Vioxx debacle."

(Zyprexa) is now known to cause extreme weight gain and concomitant diabetes, but the manufacturer hid this for years.[22] The manufacturer of paroxetine (Paxil) was charged in a massive lawsuit with hiding evidence about the harmful effects of the drug for years. These harms include withdrawal symptoms and an increase in suicidality in children and teenagers. Evidence on the harms of oseltamivir (Tamiflu) largely remain unpublished, despite the massive stockpiling of the drug by western countries in recent years.[23]

The running example of this chapter, rosiglitazone, again provides a striking illustration of the secrecy surrounding evidence of harms. In this case regulators themselves contributed to such secrecy. After several trials suggested that rosiglitazone may cause cardiovascular harms, Steve Nissen requested data from the manufacturer (GlaxoSmithKline), which refused to share the data. But due to the lawsuit mentioned above regarding paroxetine, the company had already been required to develop a registry of data from their trials. From this registry Nissen was able to access data from forty-two trials of rosiglitazone, of which only seven had been published. The resulting meta-analysis showed that rosiglitazone increases cardiovascular events by 43 percent. Within twenty-four hours of submitting the meta-analysis to the *New England Journal of Medicine*, one of the peer reviewers had faxed a copy of the article to GlaxoSmithKline. Internal emails in the company discuss the similarity of Nissen's findings to their own analysis which they had performed years earlier but had not published. Moreover, the FDA had performed their own analysis, which reached similar conclusions, but also did not publicize their findings.[24] Both the FDA and GlaxoSmithKline already had known of the cardiovascular harm caused by rosiglitazone, but neither the regulator nor the company had publicized this finding.

Indeed, regulators are not only often powerless against such secrecy, they are often complicit in it. Here is another example. Researchers doing a systematic review of the diet drugs orlistat and rimonabant tried to get unpublished data from the European Medicines Agency (EMA). The EMA rejected the request for data, by invoking protection of commercial interests and intellectual property. The researchers appealed to the European Union Ombudsperson, who found EMA to be guilty of maladministration, and found its arguments for secrecy to be unwarranted. Nevertheless, the EMA continued to withhold the evidence. Finally, the researchers were sent sixty pages. However, these sixty pages were almost entirely redacted by the EMA.[25]

---

[22] The company (Eli Lilly) "engaged in a decade-long effort to play down the health risks of Zyprexa according to hundreds of internal Lilly documents and e-mail messages among top company managers" (Berenson, 2006).

[23] The systematic bias and fraud surrounding paroxetine included millions of dollars in undisclosed payments from its manufacturer (GlaxoSmithKline) to psychiatric researchers, deliberate withholding of evidence showing that paroxetine was ineffective in children, and rampant publication bias. For a book-length exposition of this case, see (Bass, 2008). On oseltamivir, see (Doshi, 2009).

[24] In an internal email the director of research at the company wrote "FDA, Nissen, and GSK all come to comparable conclusions regarding increased risk for ischemic events, ranging from 30% to 43%!" (Harris, 2010).

[25] This case is discussed in Goldacre (2012). In 2009 rimonabant was taken off the market because it caused an increased risk of psychiatric problems and suicide.

Sometimes regulators are not complicit in such secrecy, but are simply inept. Oseltamivir again provides a striking example. When researchers set out to update their systematic review of oseltamivir in 2009, they found that the FDA post-market Adverse Event Report System had fewer entries in total than Roche's own post-market surveillance system had for just neuropsychiatric harms. The Roche system listed 2466 neuropsychiatric events between 1999 and 2007, of which 562 were classified as serious, while the FDA system only noted 1805 events of any kind (see (Doshi, 2009)).

When secrecy of evidence about harms of medical interventions is threatened by vigilant researchers, manufacturers can respond belligerently. Rosiglitazone, again, provides a good illustration. John Buse, a diabetes researcher, gave two talks arguing that rosiglitazone may have cardiovascular risks. GlaxoSmithKline executed an orchestrated campaign to silence him. This plan appears to have been initiated by the company's head of research, and even the chief executive officer was aware of it. The company referred to Buse as the 'Avandia Renegade,' and in contact with Buse and his department chair there were threats of lawsuits.[26] Buse responded to the company with a letter that asked them to "call off the dogs." Later Buse expressed embarrassment that he caved in to the pressure of GlaxoSmithKline. By 2007, the year that Nissen's meta-analysis was published, the FDA estimated that rosiglitazone had caused about 83,000 heart attacks since coming on the market in 1999.

What is wrong with secrecy of trial data? Evidence from medical research is, arguably, a public good that should be available for all to see, as clean water should be available for all to drink. Lemmens and Telfer (2012) argue that access to evidence from medical research is a fundamental component of the right to health. Secrecy of evidence from medical research impedes informed decisions (in the context of this chapter, by minimizing the apparent harms of interventions), and thereby frustrates what is arguably a fundamental right. More straightforwardly, physicians, policy-makers, and patients cannot make informed treatment decisions if they do not have access to existing evidence on the harms of medical interventions.

9.7 Conclusion

Because harms of medical interventions are systematically underestimated at all stages of medical research, policy-makers and physicians generally cannot adequately assess the benefit-harm balance of medical interventions. Many seem to think that regulators do adequately assess the benefit-harm balance of medical interventions. Unfortunately, this is far from true. David Graham, the well-known FDA epidemiologist whom I cited

---

[26] The company's head of research wrote in an internal email: "I plan to speak to Fred Sparling, his former chairman as soon as possible. I think there are two courses of action. One is to sue him for knowingly defaming our product even after we have set him straight as to the facts—the other is to launch a well planned offensive on behalf of Avandia." This passage is cited in a U.S. Senate Committee on Finance report. The full report, titled "The Intimidation of Dr. John Buse and the Diabetes Drug Avandia," was published in 2007 and is available online.

above, claimed that "Safety isn't on the radar screen regardless of what FDA officialdom would have you believe. Scientifically, FDA uses statistics in a biased manner that favors industry at the expense of patient safety" (2005). In this chapter I have argued that by the time regulators are assessing the benefit-harm profile of a medical intervention, the harms of the intervention have been systematically underestimated.

Various solutions have been proposed to address some of the problems of detecting harms of medical interventions. There are some obvious candidates, including increasing the quality of evidence in the hunt for harms, in order to increase $power_H$, and improving the accessibility of such evidence when it is available. One reason for the underestimation of harms involves a fundamental trade-off between $power_B$ and $power_H$, and too frequently the ability to detect harms of medical interventions is sacrificed in this trade-off. When evidence on harms exists, it is often publicly unavailable. The harms of medical interventions are systematically underestimated by medical research. Generally, medical interventions are more harmful than suggested by the results of all stages of medical research.

PART III

Evidence and Values

10

Bias and Fraud

10.1 Introduction

Medical research is prone to many forms of bias, and in the worst cases fraud. In the context of medical research, a bias is a property of the design of a research method, the implementation of the method, or the interpretation of the evidence generated by that method, such that the evidence is misleading with respect to the effectiveness or safety of the intervention under investigation. Biases are hindrances to truth. Moreover, biases are asymmetric with respect to truth: bias in medical research tends to generate evidence that systematically suggests that interventions are more effective and safer than they truly are. In §10.2 I survey some of the more prominent forms of bias in medical research.

Medical researchers respond to concerns about bias by introducing methodological safeguards in research. These safeguards can be relatively reliable tactics to minimize the threat from a particular form of bias. Thus, some of the biases that have introduced systematic error in evidence at some point in history do so less today. Conversely, other biases that were less prominent in earlier eras of medical research have become more prominent today. Perhaps the most worrying form of bias in medical research now is publication bias (§10.3).

The worst forms of bias are characterized by the intention of those involved to deceive others for personal or professional gain—research fraud is an extreme form of research bias in which researchers intend to introduce systematic error into the publicly available evidence. Because fraud involves deception, it is hard to estimate how ubiquitous it is. Regardless, the distinction between bias and fraud is not sharp in a context in which there are such strong financial and professional incentives at stake. In §10.4 I discuss several cases of fraud in biomedical research.

Most research on medical interventions is sponsored by the companies that manufacture the interventions, and these companies often have control over the design, execution, and reporting of the research. In itself this is not necessarily cause for concern, but research sponsored by industry is much more likely to be favorable to a company's product compared to research funded by public institutions like the NIH. The quality of studies funded by industry tends to be lower than publicly funded studies. Moreover, lower-quality studies tend to have higher estimates of the effectiveness of an experimental intervention than do higher-quality studies: the more carefully a study is designed and performed to mitigate systematic bias, the less effective interventions appear (Chapter 7). In §10.5 I articulate the various conflicts of interest at play in medical research and regulation.

These problems have received much attention by others. The point of this chapter is to provide support to one of the key premises of the master argument for medical nihilism. Given the ubiquity and variety of bias and fraud in medical research, evidence that suggests that a medical intervention is effective is likely to be published regardless of whether or not the intervention is in fact effective. Allen Frances, the head of the group that published the DSM-IV, put the point plainly (2015): "It's been many years since I have trusted anything I read in a medical or psychiatric journal." In terms of the master argument, the prior probability of typical evidence from medical research is high because of the ubiquity of biases, and thus can provide little confirmation to hypotheses regarding the effectiveness of medical interventions (§10.6).

10.2 Varieties of Bias

Bias in assessing the effectiveness of medical interventions can take a variety of forms. Throughout this book I refer to several forms of bias in medical research. In this section I describe the most prominent of these.

10.2.1 Confirmation bias

One of the most pervasive yet simplest forms of bias is confirmation bias.[1] The idea is simple: people tend to give more weight to evidence that confirms their pre-existing beliefs and less weight to evidence that disconfirms these beliefs. If a person consumes a medical intervention to treat a disease, that person (and their physicians, friends, and family) will tend to interpret his subsequent states as being caused by the intervention, whether or not such states actually were caused by the intervention. Confirmation bias is confounded by the fact that many illnesses improve on their own as part of the natural course of the illness, like the common cold. Confirmation bias is further exacerbated by the placebo effect, in which consumers of medical interventions report improvements in their health merely as a result of receiving some form of treatment—the mere consumption of a medical intervention, independent of the causal efficacy of the physically active part of the intervention (if there is such an efficacy) generates an expectation of improvement. Such expectations contribute to confirmation bias.

Consider the impact of the natural course of diseases on confirmation bias. Suppose the severity of symptoms of a disease periodically fluctuates over time (like bipolar disorder), so that there are more severe periods and less severe periods. If a person seeks treatment during more severe periods, the severity of her symptoms will

---

[1] Many kinds of biases have multiple terms; my discussion here sticks loosely to conventional terminology.

decrease after intervention, whether or not that intervention causally contributed to a decrease in symptoms. Or suppose the severity of symptoms of a disease gradually decreases over time—the same consideration applies: the severity of symptoms will decrease after intervention, whether or not that intervention causally contributed to a decrease in symptoms. So, there are two ways that the natural course of a disease can generate a correlation between the use of a medical intervention and the alleviation of symptoms. In both it is merely the passage of time that causes the alleviation of symptoms. Some diseases could have both a periodically fluctuating severity of symptoms, the overall severity of which gradually decreases over time. The consequence of this is that single case reports about experiences with medical interventions should be interpreted with great caution. This is especially so for diseases that have variable severity of symptoms.

In the context of medical research, confirmation bias has been mitigated by methodological safeguards. This includes the use of control groups, which allows for an assessment of the effectiveness of an intervention in light of the natural course of an illness, and concealment of how particular research subjects are allocated to experimental or control groups, which is sometimes called 'blinding' (this mitigates the expectation of improvement, since ideally subjects do not know if they are in the control group or not). The use of placebos is an attempt to ensure that subjects do not guess what group they are in.

These methodological safeguards are not always successful. For example, it is common for research subjects to accurately guess if they are in the experimental group or the control group, based on their responses to the experimental intervention that they have been assigned—the blinding is broken. Kirsch argues that blind-breaking is especially problematic in psychiatric research.[2] People diagnosed with psychiatric diseases tend to be very 'placebo responsive'—they report large improvements in their health when taking a placebo. The trouble is that subjects in trials can guess what group they are in based on the side effects that they experience; for example, if a patient in a trial of antidepressants begins to experience sexual dysfunction, he might accurately guess that he is in the group receiving the antidepressants rather than the group receiving placebo. If this occurs, then the subject's expectations are liable to contribute to confirmation bias.

There have been several randomized and blinded trials in which subjects were asked (after the trial was completed) to guess their group, and the results consistently show that guessing one's group assignment is greater than chance, and can be as high as 90 percent. That is, blind-breaking is regularly occurring. This is sometimes called detection bias. Detection bias is deeply troubling, given the impact of confirmation bias. Unfortunately, most trials do not at present ask about group guessing, and so we do not know the extent of detection bias. Asking about group guessing would be a simple addition to the design of most trials and would provide another parameter

[2] (Kirsch, 2011). See also (Holman, 2015).

with which one could assess the quality of trials to supplement the parameters discussed in Chapter 7.

Despite their imperfect implementation, these methodological safeguards—control groups, concealment of group allocation, and use of placebos—are important innovations to mitigate confirmation bias. However, I have dwelled on confirmation bias here because, despite these methodological safeguards, confirmation bias still plays a prominent role in biasing our views about the effectiveness of medical interventions. As we have seen, within medical research the safeguard of subject allocation concealment often fails, which can lead to confirmation bias. But worse and much more widespread, outside the context of controlled medical research there are no such safeguards, and most people reason in ways that are influenced by confirmation bias. It is very common for people to attribute improvements in their health to medical interventions that they have recently used. This is not always unreasonable. But very often it is unreasonable, because of the threat of confirmation bias, and its exaggeration by the natural course of diseases and enhancement by the placebo effect. Social media and traditional media are often used to promulgate personal anecdotes that make claims about the effectiveness of particular medical interventions, thereby greatly increasing the pernicious epistemic consequences of confirmation bias.

10.2.2 Design bias

Another widespread form of bias in medical research is design bias. Medical studies that employ some of the safeguards noted above can nevertheless have design features—often quite subtle properties of a study—that render the study biased. A design bias is a bias built into the design of a research method.

A widespread type of design bias is known as selection bias. One of the basic elements of medical research, as noted above, is to compare subjects in one group—those that receive the experimental intervention—with subjects in a control group—those that receive a placebo or competitor intervention. To reliably infer that the intervention causes the outcome in question, the subjects in the two groups must be similar with respect to all the other possible factors that can influence the probability of the outcome.[3] Selection bias occurs when there is a salient difference between the subjects in the experimental group and the subjects in the control group brought about intentionally or unintentionally by researchers or subjects themselves.

Randomizing the allocation of subjects into groups is an attempt to mitigate selection bias, by taking the allocation of subjects out of the control of researchers. As Worrall (2002) and others argue, random allocation of subjects does not guarantee that subjects in the experimental group will be similar in all salient respects to subjects in the control group, but the hope is that randomization will at least minimize such differences.

---

[3] For a canonical formal statement of this requirement, see (Cartwright, 1979). See also (Pearl, 2009) and (Fuller, forthcoming).

Selection bias is an especially salient concern for non-randomized studies such as case-control studies; however, Worrall and other critics are correct to note that even trials that employ safeguards such as randomization to avoid selection bias can still have experimental groups that are imbalanced with respect to confounding factors, because real trials are not ideal trials (see Chapter 5).

In Chapter 8 I noted the importance of instruments for measuring effectiveness, and I described ways in which instrument bias can introduce error into this measurement. Instrument bias is a form of design bias in which the measuring instruments employed in a trial are tuned in such a way that a medical intervention appears more effective and safer than it truly is. Recall my discussion of the HAMD scale in Chapters 8 and 9. This scale is designed in such a way that trials that employ the scale to test experimental antidepressants generate evidence that makes such antidepressants appear to be more effective and safer than they truly are.

Here is another example of design bias from the field of toxicology. Bisphenol A (BPA) is used in some plastic products and has been related to cancer and other harmful effects, due to its chemical similarity to estrogen. In studies testing the harms of BPA in rats, 90 percent of non-industry studies reported harmful effects of low dose BPA, but 0 percent of industry studies did so (Wilholt, 2009). A possible explanation of this is that industry studies more often used strains of rats that were less sensitive to estrogen, thereby introducing bias in their experimental design via the choice of animal model.

In short, randomized trials have safeguards to minimize some of the most ubiquitous forms of bias, including confirmation bias and selection bias. But as my discussion in Chapters 5, 8, and 9 showed, even randomized trials can have many biases built into their design. Instrument bias, a type of design bias, is an example of this. Recruitment bias is another example. Thus, even if a trial successfully employs methodological safeguards to mitigate confirmation bias and selection bias, other forms of design bias such as instrument bias can hamper the trial's veracity.

10.2.3 Analysis bias

Once a trial is completed and data is gathered, how that data is analyzed can be subject to bias. One form of analysis bias occurs when researchers submit their data to multiple analyses in an overzealous hunt for a statistically significant difference between subjects in the experimental group and subjects in the control group. A trial can gather data on multiple outcomes (measured parameters of subjects), data can be grouped in a potentially infinite number of ways (for example, subjects' ages can be measured to the day, but can then be binned by month-long intervals, year-long intervals, five-year intervals, and so on), and a variety of statistical tests can be applied to a single set of data. The problem is that spurious statistical relationships very likely will be detected when researchers can perform so many possible analyses. For example, suppose that in a particular set of data a spurious finding would occur in 5 percent of analyses, and researchers perform one hundred analyses—in this scenario, one would expect to observe roughly five spurious statistical relationships. A term that is used to describe this practice is p-hacking.

Consider the many subjective decisions that must be made by data analysts, including how missing data and outlying data are handled, how continuous data are binned, or how skewed distributional data is treated (for instance, sometimes skewed data is transformed into a 'normal' distribution because some statistical analyses require normal distributions). Analysts have wide latitude in such decisions, and every uncon-strained analytic decision permits more possible analyses with more potential for generating spurious results.

An interesting study highlighted the role of design bias and analysis bias in overesti-mating the effectiveness of medical interventions. Kaplan and Irvin (2015) found all publications of randomized trials that had been funded by the National Heart, Lung, and Blood Institute (NHLBI) from 1970 to 2012 that were about interventions intended to decrease the risk of cardiovascular disease and death. They split the trials into two groups: those published prior to the year 2000, and those published after. Although this was a small group of trials (thirty in the pre-2000 group and twenty-five in the post-2000 group), it was an important group of trials because they were publicly funded, and so there was less control over the trials on the part of industry (although many of the trials were partly financed by industry). The year 2000 was special, because after 2000 all of these trials were pre-registered in a database of trials, but prior to 2000 none were. The results were striking. Prior to 2000, 57 percent of trials suggested that the interventions under investigation had a benefit, while after 2000 only 8 percent of trials did. Also interesting was that prior to the year 2000, the various trials showed a wide variability in reported effect sizes, while after 2000 the trials showed very low variability of effect sizes with the risk difference clustering around 0. In other words, after 2000 on average the interventions did not help the subjects in the trials.

Kaplan and Irvin (2015) note that after 2000 there was strong NHLBI involvement in the design and execution of trials, whereas before 2000 there was less NHLBI involvement and more industry involvement.[4] After 2000 such design bias was miti-gated. They argue that prior to the year 2000 trial designers "had a greater opportunity to measure a range of variables and to select the most successful outcomes when reporting their results," and they could freely modify their statistical analyses post hoc in a search to find any statistically significant results. Once trials had to be pre-registered such analysis bias was mitigated.[5]

I have made no attempt to be thorough in articulating all the types and subtypes of biases in medical research. This would involve a massive undertaking. For example, David Sackett, one of the founders of the evidence-based medicine movement,

---

[4] In the words of Kaplan and Irvin (2015): "It is possible that industry conducts trials designed to demonstrate effectiveness while NHLBI uses its resources when there is true equipoise."

[5] Every single treatment that had been the object of investigation in these trials had already been approved by regulators, despite the fact that the most carefully designed and analyzed trials showed that the vast majority of these treatments are useless.

described fifty-six subtypes of bias in medical research, including: volunteer bias, missing clinical data bias, withdrawal bias, compliance bias, bogus control bias, exposure suspicion bias, recall bias, instrument bias, and repeated peeks bias (1979). Medical research can succumb to many forms of bias, and I have noted only some of the more prominent ones here.

There is a large second-order empirical literature that shows a strong influence between the source of funding of medical research and the results of that research. For example, one research group evaluated 192 randomized trials that compared statins against other statins or non-statin alternatives.[6] This group found widespread biases in the trials (including blind-breaking). Moreover, they found that if a drug company sponsored a trial, the conclusions reported from that trial were thirty-five times more likely to favor the drug made by the sponsoring company. A variety of kinds of conflicts of interest, discussed in §10.5, can generate and exacerbate the biases discussed in this section and the next.

10.3 Publication Bias

Once a study has been completed and analyzed, researchers and their company sponsors can choose whether or not to publish the results. Publication is systemically biased. Evidence that suggests that a medical intervention is ineffective or harmful is less likely to be published than evidence that suggests that a medical intervention is effective and safe. This ubiquitous phenomenon is known as publication bias. Publication bias plays a prominent role in the arguments of Chapters 6 and 9.

A form of bias related to publication bias is reporting bias. This occurs when researchers do in fact publish results of a trial, but publish only some of the data that they had gathered. Reporting bias and publication bias both amount to withholding evidence about medical interventions.

Evidence for the alleged effectiveness of oseltamivir (Tamiflu) provides a striking example of publication bias. Oseltamivir was sold in vast quantities as a treatment for influenza. As of 2012, 60 percent of data from phase 3 trials had not been published. A group of researchers performed a careful review on oseltamivir, and found that only five trials could be included, because of publication bias and trial design shortcomings, and even these trials had some bias (though recall my concern in Chapter 6 about such filtering when performing a systematic review). These researchers contacted the lead authors of the five trials; only three responded, and all claimed that they did not have the original data and referred the researchers to the manufacturer of oseltamivir.[7] The manufacturer did not provide the data. The data that the researchers did have access to,

---

[6] This was reported in (Bero et al., 2007)—there have been many examples of these sorts of findings published in recent years.

[7] Reported in (Jefferson, Jones, Doshi, & Del Mar, 2009) and (Jefferson et al., 2012). In addition to publication bias, the latter article noted widespread design bias in the relevant trials.

despite the rampant publication bias, showed no beneficial effect of oseltamivir on prophylaxis against influenza, no beneficial effect on complications from influenza, no beneficial effect on viral transmission, and only very modest benefit on shortening the duration of symptoms.

A demonstration of the ubiquity of publication bias and reporting bias was provided by a German health technology assessment agency, which analyzed 101 trials for sixteen drugs that had been evaluated by the agency over a recent five-year period.[8] Of the 101 trials, thirty-six had no corresponding publication. Among all the trials there were 1080 outcomes measured, but in the corresponding publications there was data presented on only 250 of the outcomes. In other words, 77 percent of measured outcomes had not been reported in publications. Most of the drugs under investigation in these trials were antidepressants, but the drugs also included interventions for diabetes (including rosiglitazone and pioglitazone, discussed in Chapter 9), asthma, stroke, and Alzheimer's disease.

Some cases of publication bias amount to fraud. The case of paroxetine (Paxil), discussed in the following section, is an example.

10.4 Fraud

Fraud in research involves deliberate deception with the aim of personal or professional gain. There are cases of isolated researchers engaged in outright fraud (such as the case of a researcher who fabricated data for twenty-one studies of painkillers), and there are cases that involve multiple members of a large institution coordinating their activities to systematically deceive the public, regulators, and physicians. Cases of outright fraud are probably rare in medical research, though because fraud involves an intention to deceive it is difficult to estimate how frequently it occurs. I illustrate fraud with a well-known case: the marketing of paroxetine for children under the age of eighteen by GlaxoSmithKline. Nevertheless, I do not dwell much on fraud here, because in my view the greatest threat to assessing the true safety and effectiveness of medical interventions involves the (intentional or unintentional) exploitation of the subtler threats to truth discussed in §10.2 and straightforward publication bias discussed in §10.3. Indeed, the example discussed here demonstrates how nefarious publication bias can be, and that in the context of medical research, publication bias ought to be (and sometimes is) considered fraudulent.

The antidepressant paroxetine (Paxil, Seroxat) was not approved for patients under the age of eighteen, but despite this GlaxoSmithKline promoted its use in children and teenagers. GlaxoSmithKline conducted three trials of paroxetine for treating depression in patients under the age of eighteen, and none of these trials showed any benefit. Data from one of these trials suggested that teenagers on paroxetine were more likely to

---

[8] This was reported in (Wieseler et al., 2013). The English name of the agency is the Institute for Quality and Efficiency in Health Care.

attempt suicide than teenagers on placebo. The corresponding publication downplayed this finding, and falsely claimed that paroxetine was an effective treatment for childhood depression. The company used this article to promote the use of paroxetine in children. Because of its fraudulent actions involving paroxetine and other drugs—bupropion (Wellbutrin), another antidepressant, and rosiglitazone, the drug for type 2 diabetes discussed in Chapter 9—GlaxoSmithKline was forced to pay a $3 billion criminal and civil settlement (which at the time was the largest fine ever demanded of a pharmaceutical company).

As suggested above, it is difficult to estimate the frequency of fraud in medical research, because the very nature of fraud involves an intention to deceive, but my suspicion is that straightforward cases of data fabrication are rare, while more subtle practices of deception that involve exploiting the many possible biases in medical research are ubiquitous.

10.5 Conflicts of Interest

Biases and fraud in medical research can be motivated and exacerbated by conflicts of interest. Conflicts of interest in medical research are ubiquitous (Resnik, 2007). Conflicts of interest exist among industrial manufacturers, medical researchers, regulatory agencies and their advising committees, clinical guideline developers, patient advocacy committees, medical educators and textbook authors, and even medical journals. Perhaps the most important form of conflict of interest is financial, in which there is a risk of compromised judgment, intentionally or unintentionally based on financial considerations. In this section I provide several examples of conflicts of interest in medical research and regulation to illustrate how significant the problem is.

Medical researchers usually have their primary employment in universities and hospitals, but in addition many hold executive positions in private companies that sponsor research, own shares of corporations that manufacture pharmaceuticals, receive payment for consulting with industry, own patents that are licensed to industry, and receive fees for enrolling their patients in trials. Three examples discussed by Angell (2009) suffice to illustrate this problem. Joseph Biederman is a professor of pediatric psychiatry at Harvard who received $1.6 million in consulting and speaking fees from pharmaceutical companies that manufacture drugs that he promotes (notably, for alleged childhood bipolar disorder). Alan Schatzberg, who has been chair of the psychiatry department at Stanford and president of the American Psychiatric Association, owned about $5 million worth of stock in a company that was testing the drug mifepristone as a treatment for depression, and at the same time Schatzberg was the lead investigator for a NIMH grant on that very same drug. Charles Nemeroff, who was chair of the psychiatry department at Emory University, received a $4 million grant from NIMH to study drugs made by GlaxoSmithKline—Nemeroff was required to disclose financial relations to NIMH, yet Nemeroff did not disclose about $500,000 he received from GlaxoSmithKline. From 2000 to 2007 Nemeroff was paid over $2.8 million from pharmaceutical companies, and he failed to report $1.2 million to his university.[9]

As Resnik (2007) notes, conflicts of interest are present not just among individual researchers, but also among institutions. An obvious form of this is the financial incentive for pharmaceutical companies to demonstrate that one of their patented products is effective. Even medical journals can have conflicts of interest, because many journals receive a large portion of their budgets from advertisements paid for by manufacturers of the products that are evaluated in articles in those journals.[10]

Financial conflicts of interest exist in regulatory agencies. For example, Resnik (2007) and Krimsky (2003) discuss an investigation into conflicts of interest at the FDA. The FDA relies on advisory committees, which are composed of internal staff and external consultants. The investigation examined 159 meetings by eighteen advisory committees, and found that there was at least one committee member with a financial conflict of interest in 146 of the meetings, and over half the committee members in eighty-eight of the meetings had financial interests that were immediately relevant to the interventions being evaluated by the committee. That is, a majority of committee members in the majority of committee meetings of the FDA had financial conflicts of interest. Moreover, the FDA collects fees from companies when the companies submit applications for drug approval, and the majority of those fees go toward staff salaries.

Two common tactics to mitigate the pernicious influence of conflicts of interest are: constraint, in which financial relationships are capped at certain values, and disclosure, in which financial relationships are required to be disclosed. For an example of constraining conflicts of interest, some universities limit the amount that their staff can receive from private companies. Many granting agencies and journals require scientists to disclose the various sources of funds they have received from private companies. It is a platitude that these tactics are useful for mitigating the influence of conflicts of interest on bias in research. But these tactics are hardly sufficient. A stronger tactic would be full prohibition of conflicts of interest, enforced by universities, professional societies, journals, or more powerful institutions. Resnik (2007) argues against full prohibition of conflicts of interest by claiming that scientific progress would be hampered if such a policy were instituted; for example, if a journal prohibited conflicts of interest, Resnik claims, then scientists might be deterred from publishing. This empirical claim about the motives of scientists is dubious—scientists have typically sought to publish their work for motives other than financial reasons, such as a desire to share their work in a communal enterprise of discovery, or a desire for credit. An even stronger strategy would be to altogether eliminate financial incentives

---

[9] Reported by Harris (2008). Incidentally, Nemeroff and Schatzberg are coeditors of an important textbook in psychiatric pharmacology, and Nemeroff was editor-in-chief of the journal *Neuropsychopharmacology*.

[10] Professional associations are also liable to conflicts of interest—in 2006, for example, the American Psychiatry Association received about 30 percent of its $62.5 million budget from pharmaceutical companies (Carey & Harris, 2008).

in medical research, by a variety of means such as eliminating intellectual property protection—I discuss this more radical idea in Chapter 12.

10.6 Formalizing Bias

In this section I articulate the influence of bias on medical research in formal terms, in anticipation of the master argument for medical nihilism presented in Chapter 11. Evidence that is generated from methodologically weak studies (methods that are biased) should be assigned a high prior probability. This is true regardless of whether or not the evidence provides some support for a hypothesis of effectiveness. That is because a methodological flaw permits a possible explanation for the observed evidence that is an alternative to the hypothesis of interest. In other words, given the presence of methodological biases, there are alternative hypotheses, which can explain the apparent evidence that an intervention is effective other than the hypothesis that the intervention is in fact effective, and when this is the case, the prior probability of the evidence increases. And evidence with a higher prior probability is less confirming (see Appendix 1). This is a convoluted way of stating the obvious: more biased methods produce evidence that is not as confirming as less biased methods. Stating this obvious point in formal term will be important for the master argument in Chapter 11, so here I belabor the point with care.

One can see the influence of biases on the prior probability of the evidence more clearly upon consideration of the principle of total probability (Appendix 1). First I state the principle in abstract terms, and then I apply it to the confirmation of hypotheses by biased methods. The principle of total probability holds that if $Z_1 \ldots Z_n$ is a partition of possible states or events (that is, if $Z_1 \ldots Z_n$ are distinct states or events, which together make up the entire set of possible states or events), then for any event Y the probability of Y is:

$$P(Y) = P(Y|Z_1)P(Z_1) + P(Y|Z_2)P(Z_2)\ldots + P(Y|Z_n)P(Z_n)$$

Suppose that we have some evidence (E) which suggests that an intervention is effective, and one possible explanation of this evidence is the hypothesis that the intervention is in fact effective (call this H_X). E provides at least some confirmation to H_X. But how much? An alternative hypothesis that could explain E is that the intervention is not in fact effective and that E occurred by chance (H_C). Moreover, if the trial that generated E suffered from i distinct forms of bias, then other alternative hypotheses are that the intervention is not in fact effective and that E was the result of one or more of the biases (call these alternative hypotheses H_{Bi}, one for each of i biases). Suppose that these are all the possible hypotheses, and suppose for the sake of simplicity that the various hypotheses are mutually exclusive (this is of course an idealization—dropping this presupposition would not change the general point but would complicate its presentation). Drawing on the principle of total probability:

$$P(E) = P(E|H_X)P(H_X) + P(E|H_C)P(H_C) + P(E|H_{B1})P(H_{B1})$$
$$+ P(E|H_{B2})P(H_{B2}) \ldots + P(E|H_{Bi})P(H_{Bi})$$

The terms $P(E|H_n)$ are the likelihoods, and can be understood as representing how well any particular hypothesis explains E. Notice that as the number of biases increases, P(E) increases (until it reaches its maximum of 1), because each bias adds an alternative hypothesis and thus adds a positive term $P(E|H_{Bi})P(H_{Bi})$ to the right side of the equation. Notice also that as the prior probability of any hypothesis increases, including any of the alternative hypotheses H_{Bi}, P(E) increases; thus, the more likely it is that a method suffers from any particular bias, P(E) increases. Finally, notice that the more any particular H_{Bi} can explain E, P(E) increases.[11] Since, by Bayes' Theorem, confirmation of a hypothesis is inversely related to P(E) (see Appendix 1), the more possible biases, and the higher the prior probability of any particular bias in the method that generates E, and the more any particular bias can explain E, the less confirmation a hypothesis of effectiveness receives from evidence that seemingly supports that hypothesis.

10.7 Discussion

Even if a trial employs particular methodological safeguards against certain prominent forms of bias, this does not entail that the evidence from the trial is reliable. There is a view that trial design principles such as random allocation of subjects to experimental groups mitigate biases, and so evidence from trials that employ such principles is compelling. This is based on the view that randomized trials are the most reliable method in medical research, and meta-analyses of such trials are the final word on whether or not a medical intervention is effective. But given the great plurality of possible biases in medical research, from trial design to reporting and analysis and publishing, it does not follow that simply because a trial minimized one set of biases it therefore eliminated all biases and is therefore reliable (see Chapters 5, 8, and 9).

One result of the pervasiveness of biases and the nefarious conflicts of interests within medical research is that one ought to place less confidence in medical research and the clinical guidelines that are based on such research. Marcia Angell, the former editor of *The New England Journal of Medicine*, puts this point as follows: "it is simply

---

[11] One could formulate the impact of bias by a likelihood analysis without relying on prior probabilities or the principle of total probability. Likelihoodism, defended by Sober (2008) and Royall (1997), among others, holds that E supports H_X over some competitor hypothesis H_B if and only if $P(E|H_X)/P(E|H_B) = w > 1$. The greater the number and severity of biases, the more that H_B explains E and so the higher is $P(E|H_B)$ and thus the lower is w, and thus the less likely it is that E confirms H_X. Though this approach has the virtue of not relying on priors or the principle of total probability, it is less informative regarding how biases support the master argument for medical nihilism (Chapter 11).

no longer possible to believe much of the clinical research that is published, or to rely on the judgment of trusted physicians or authoritative medical guidelines" (2009).

One possible objection to the general thesis argued for in this book—that one should have little confidence in the effectiveness of medical interventions—could be motivated by appealing to personal anecdote. For example, one often hears that a certain drug was effective for a particular person. In the present era of mass communication facilitated by the internet, such anecdotes are ubiquitous. The trouble with this sort of response to medical nihilism is that personal anecdotes are shot through with bias. The key bias at play in personal anecdotes is confirmation bias, compounded by the fact that many diseases have a natural course of improvement even in the absence of an intervention. For this reason, the evidence-based medicine community long ago learned to place little weight on anecdotal evidence from physicians regarding interventions that they had employed in their practice. When assessing the general plausibility of medical nihilism one ought to place very little weight on counterexamples that are based on personal or professional anecdote.

A practice in medical research that has generated much critical commentary is ghostwriting, in which the design, performance, and analysis of a trial, and the sub-sequent writing of the corresponding article, are done by private companies under contract with a drug manufacturer, and then the names of high-profile physicians who had little or nothing to do with the trial are listed as authors of the corresponding article. In itself it is not an obvious transgression of contemporary scientific norms; as Kukla (2012) argues, "the entrenched organizational practices and architecture of contemporary biomedical research have rendered much of it authorless in ways that raise serious challenges for social epistemology." Medical research today is collabora-tive, decentralized, and interdisciplinary—Kukla argues that medical research is not well suited to the traditional notion of authorship, in which one can identify particular individuals who are responsible for a text. Authorship is attributed according to insti-tutional status rather than labor.[12]

Kukla notes that this raises epistemic problems for medical research that exacerbate the problems introduced by bias in this chapter. Many philosophers of science hold that non-epistemic values can influence scientific reasoning. The key argument for this is based on the notion of inductive risk: since a hypothesis can be wrongly accepted as true when it is false or wrongly accepted as false when it is true, and since these errors can have practical consequences, the way one ought to balance strategies to avoid these errors is influenced by values pertaining to those consequences.[13] Because there is so much malleability in medical research, there are many points in medical research at which such values can intervene. And since medical research is so radically distributed,

---

[12] As Kukla puts it, "many contemporary biomedical research articles, even when they accord with current norms of scientific practice, have no author at all in any traditional sense," and that is the case whether or not the articles are ghostwritten.

[13] See (Brown, 2013), (Douglas, 2000), (Elliott, 2011), and (Rudner, 1953). In Chapter 12 I discuss inductive risk in the context of drug regulation.

Kukla argues that the values that influence medical science are themselves distributed and perhaps inarticulable. However, for most trials, their design, implementation, analysis, and reporting may be controlled by a centralized authority—a handful of corporate scientists and executives, say—in such cases, contrary to Kukla's thesis, there is an identifiable locus at which non-epistemic values can have a pernicious epistemic influence, and such influence can be articulable (for examples see Chapter 9 and §10.4).

An obvious response to the concern about biases in medical research is that trial design, implementation, analysis, and reporting should be improved to mitigate the influence of these biases. If this were to be done—which of course it should—then that would decrease the plausibility of one of the premises for medical nihilism, but it would increase the plausibility of another. The review by Kaplan and Irvin (2015) discussed above is one of many empirical demonstrations that show that better research methods in medicine lead to lower estimates of effectiveness. The more that biases are controlled for in a method, the more that the evidence from that method is truth-conducive. Better methods increase the veridicality of evidence, and greater veridicality of evidence leads to lower (and more accurate) estimates of effectiveness of interventions. All empirical methods in science have some risk of error, including the best trials. Thus, if the inverse relationship between quality of research methods and estimates of effectiveness holds across the full spectrum of method quality (from totally misleading to the unattainable ideal of totally veridical), and if we further improve upon our very best research methods, we will observe even lower estimates of effectiveness. Since the best estimates of effectiveness of many medical interventions are already so low (see §11.3), and better methods would render these estimates even lower, we have another consideration that supports the general thesis of medical nihilism.

Biases in medical research support medical nihilism because the presence of such biases renders medical research less reliable, and thus the evidence from such research is less veridical, and thus the hypotheses of effectiveness that are allegedly supported by such evidence are less plausible. To articulate this argument with care, I formalized it using the tools of probability theory. This affords the application of the conclusion of the present argument, based on a concern about biases in medical research, to be employed in support of the master argument for medical nihilism, to which I now turn.

11

Medical Nihilism

11.1 Introduction

We should have little confidence in the effectiveness of medical interventions. This view, called therapeutic nihilism in the nineteenth century, has been popular at various times throughout history. Hippocrates, Montaigne, and Nietzsche held that the medical interventions of their times should not be trusted. The nineteenth century saw medical nihilism become a mainstream view within medicine itself. After the confidence about medical progress that attended the great success of antibiotics and insulin from the 1920s to the 1950s, the 1960s and 1970s saw a resurgence of medical nihilism (see my historical survey of medical nihilism in Chapter 1). Today the view has mostly fallen out of favor, and there is widespread confidence in the effectiveness of medical interventions. In contrast to such confidence, this book offers a series of arguments that support contemporary medical nihilism.

The arguments for medical nihilism are based on a wide range of conceptual, methodological, and empirical considerations. The arguments from Part I of this book are conceptual: this analysis articulated a standard for what a medical intervention must do to be effective. Many medical interventions do not satisfy this standard. Moreover, despite some examples of magic bullets discovered in the twentieth century, we have very few magic bullets today. The arguments from Part II are methodological: medical research has developed sophisticated methods, such as randomized trials and meta-analyses, in an attempt to find interventions that are effective, but these methods are malleable. Given the great incentives to find interventions that at least appear effective, these methods are systematically biased to overestimate the effectiveness of interventions, and underestimate the harms. Part III begins with a description of these biases. In this present chapter I describe empirical findings that support medical nihilism. Many interventions have been removed from clinical practice because after they were used outside the experimental setting they appeared to do more harm than good (§11.2). Despite the great effort to find effective medical interventions, many of the most commonly employed interventions have tiny effect sizes (§11.3). All of these arguments come together in service of the master argument for medical nihilism (§11.4).

Some of the problems that I have described in this book have been voiced by medical researchers.[1] In Chapter 1 I noted several examples of prominent physicians and

---

[1] Here, for instance, is the GRADE working group, an influential group of methodologists: "Recent years have seen an increased awareness of a number of factors that influence our confidence in our estimates

epidemiologists who have described problems with medical research that lend support to medical nihilism. The conclusion of the master argument will be less surprising to those who are familiar with the details of contemporary medical research.

I articulate the master argument using a formal structure based on our best theory of scientific inference, with the aim of unifying the disparate chapter-level arguments. One's confidence in the effectiveness of a medical intervention can be represented as a conditional probability, $P(H|E)$, where H is a hypothesis regarding the effectiveness of the intervention, and E is our available evidence relevant to the hypothesis. Based on the above conceptual, methodological, and empirical considerations, a compelling case can be made that, on average, we ought to have a low prior probability in H, $P(H)$; that when presented with E we ought to have a low estimation of the likelihood of that evidence, $P(E|H)$; and similarly, that we ought to have a high prior probability of E, $P(E)$. What follows, given Bayes' Theorem, is that even when presented with evidence for a hypothesis regarding the effectiveness of a medical intervention, on average we ought to have a low posterior probability in that hypothesis, $P(H|E)$—in short, medical nihilism is compelling. The master argument is valid, because it simply takes the form of a deductive theorem; the conclusions of earlier chapters, plus empirical material in this chapter, provide warrant for the premises of the master argument.

11.2 Rejected Medical Interventions

Many medical interventions that appear promising when first introduced, and occasionally become entrenched in medical practice, are later deemed useless or even harmful, or remain in practice as relatively ineffective but best available treatment for a disease. This is clear when considering the long history of rejected interventions in medicine. But it is also apparent in recent medicine. The upshot of the ubiquity of past rejected medical interventions is that when an intervention is introduced we ought to maintain skepticism regarding its effectiveness, and such skepticism should be informed by the past frequency of rejected interventions. That is, for a hypothesis (H) regarding the effectiveness of a medical intervention, prior to considering any evidence for or against the hypothesis we ought to hold a low probability of the truth of H. In Bayesian terms, the prior probability of H, $P(H)$, ought to be low.

This historical consideration complements the argument from Chapter 4 that explained why magic bullets are rare. There have been a couple of great examples of magic bullets, including insulin and penicillin, but because of the complexity of the pathophysiological basis of most diseases and the cascading complexity of how exogenous interventions operate in human physiology, there have been few magic

---

of risk and benefit, such as poor quality of planning and implementation of the available randomized controlled trials suggesting high likelihood of bias; inconsistency of results; indirectness of evidence; and sparse evidence" (text available online).

bullets discovered in recent decades. This consideration provides further reason for our prior confidence in an effectiveness hypothesis to be low.

There are many examples of medical interventions that were once widely employed and assumed to be effective but have since been rejected as useless or even harmful. It is easy to find such historical examples: bloodletting, guaiacum (for syphilis), mercury (for syphilis and nearly everything else), hydropathy, tartar emetic, strychnine, opium, jalap, Daffy's Elixir, Turlington's Balsam of Life, Elixir sulfanilamide...[2] Once one starts looking, there is practically no end to the list of rejected medical interventions.

A survey of some of the most prominent rejected medical interventions in history is given by Wootton (2006), who argues that from antiquity until the end of the nineteenth century medicine had done more harm than good. The turning point, argues Wootton, was the advent of more careful empirical testing of medical interventions. Such narratives are a genre among those in evidence-based medicine: notice how bad medical interventions were, until we used randomized trials and got evidence about their ineffectiveness. Although the criticism of historical medical interventions is obviously consistent with my thesis (and indeed, is part of the basis for holding a low prior for any hypothesis regarding the effectiveness of an intervention), it is not the case that contemporary methodology has shown present-day medical interventions, on the whole, to be much more effective than past interventions. There are wonderful exceptions but on average this is not the case. Improved methods tend to show that interventions are less effective, both in the early days of evidence-based medicine and today (see Chapters 7 and 10).

One might appeal to this historical turning point in research methodology to argue that the motley mix of rejected interventions throughout history is the wrong reference class for assessing the probability of the effectiveness of novel medical interventions, because medicine is so much better today than it was in the past. To avoid begging the question, this response must identify features of contemporary medicine that render only interventions introduced into medicine recently as the proper reference class. Such a response faces a burden of identifying a point in time that separates inappropriately old medical interventions from contemporary medical interventions to properly demarcate the correct reference class for assessing novel medical interventions. Something about medicine from, say, the last seven or eight decades must be identified as a reason for privileging the interventions from these decades as the appropriate reference class from which we ought to assess P(H).

As suggested above, a standard feature that is appealed to in order to differentiate contemporary medicine from historical medicine is that today medicine is said to be evidence-based. Moreover, the science undergirding contemporary medicine is much

---

[2] Ridiculing past therapies has been a literary sport. In Molière's *The Imaginary Invalid* a student is examined for a physician's license; when asked how to treat dropsy, the student responded: Give an enema, then bleed the patient, and purge. When asked how to treat asthma, the answer was: Give an enema, then bleed the patient, and purge. When asked to treat a headache: Give an enema, then bleed the patient, and purge. If that treatment is at first ineffective? Give an enema, then bleed the patient, and purge.

more advanced than it was in the days of these past interventions. Today, to use the terms of the trade, we have rational drug design and evidence-based medicine, with the promise of development from research programs like genomics, proteomics, and precision medicine. To say that we ought to hold a low probability in the effectiveness of medical interventions simply because so many past interventions have been rejected is unwarranted, one might respond. The right reference class for assessing the prior probability that a medical intervention will be effective—so this response goes—is the set of medical interventions introduced, say, since the 1940s or 1950s. Even if one grants that that this line can be drawn non-arbitrarily—say, by appealing to theoretical developments such as germ theory in the nineteenth century, institutional developments such as the birth of the FDA in the 1920s, or methodological developments such as the rise of randomized trials in the 1950s—the empirical case for the effectiveness of more recent medical interventions is not much stronger than the case for the effectiveness of past interventions.

Many medical interventions widely employed in the twentieth century have been rejected after their sanction for clinical use for the simple reason that sometime after they were approved for use they were later deemed harmful. By 'rejected' here I mean withdrawn from the market either under regulatory stipulation or voluntarily, due to an assessment that the intervention had an unfavorable benefit-harm profile. Some famous examples of such interventions are thalidomide, methaqualone, and lysergic acid diethylamide (otherwise known as LSD). Others include some antiarrhythmics, diethylstilbestrol, and hormone replacement therapy for menopausal women. Particular pharmaceuticals that have been removed from the market by manufacturers, prohibited by regulators, or clinically constrained or commercially marginalized due to excessive harms include isotretinoin (Accutane), rosiglitazone (Avandia), valdecoxib (Bextra), fenfluramine (Pondimin), gatifloxacin (Gatiflo), aprotinin (Trasylol), tegaserod (Zelnorm), sibutramine (Reductil), rofecoxib (Vioxx), cerivastatin (Baycol), rapacuronium (Raplon), gemtuzumab ozogamicin (Mylotarg), efalizumab (Raptiva), furazolidone (Furoxone), nefazodone (Serzone), and rimonabant (Acomplia). This latter list is only a small subset of rejected medical interventions, and includes only those rejected in the last fifteen years.[3] Many other medical interventions have not yet been prohibited but are currently suspected of doing more harm than good, and some are the subject of class action lawsuits, or are labeled with 'black box warnings,' including celecoxib (Celebrex), alendronate (Fosamax), telithromycin (Ketek), transdermal estrogen and progestin (Ortho Evra patch), risperidone (Risperdal), and olanzapine (Zyprexa).[4]

---

[3] I have not made any attempt to be comprehensive with this list.

[4] The latter has been the subject of great controversy: because Eli Lilly, the manufacturer of Zyprexa, engaged in a multi-year campaign to deceive the public about the harms of Zyprexa (it causes extreme weight gain), the company was required to pay several billion dollars in civil settlements and criminal fines. Bextra was also the subject of one of the largest criminal fines ever imposed in the United States ($1.2 billion). See Chapter 9.

Many other medical interventions widely employed in the twentieth century have been rejected because they were later deemed ineffective. For instance, in 2011 a large trial testing nesiritide (Natrecor) found no significant benefit for the symptom it was thought to help.[5] To indicate the depth of the rejection phenomenon, consider just the narrow class of nonselective monoamine oxidase inhibitor antidepressants—though these drugs were widely used in the 1950s and 1960s, they have been largely rejected, including: benmoxin (Neuralex), caroxazone (Surodil), iproclozide (Sursum), iproniazid (Marsilid), mebanazine (Actomol), metfendrazine (MO-482), nialamide (Niamid), octamoxin (Ximaol), pheniprazine (Catron), phenoxypropazine (Drazine), pivalylbenzhydrazine (Tersavid), and safrazine (Safra). Or consider another particular class of drugs, the modulators of peroxisome proliferator-activated receptors (PPAR modulators, discussed in Chapter 9). Several drugs in this class have been withdrawn by various regulators in Europe or America, including troglitazone, pioglitazone, and muraglitazar, because they cause harms such as liver damage, cancer, heart attacks, strokes, and death.

The empirical basis of this argument is only suggestive. It would be virtually impossible to determine a clear numerator and denominator of drugs from which we could determine a ratio of effective and safe drugs to ineffective or harmful drugs. However, many of the interventions mentioned here and elsewhere in this book have been among the most prominent pharmaceuticals in the world. For example, over 80 million people took rofecoxib (Vioxx), approved in 1999 but later withdrawn because it caused an increased risk of heart attack and stroke (see Chapter 9). Or consider olanzapine (Zyprexa)—sales of Zyprexa were $4.7 billion dollars in 2008. The interventions discussed here are among the most prescribed medical interventions. Thus, despite the fact that this empirical phenomenon lacks a quantitative analysis, the fact that medical interventions throughout the history of medicine and blockbuster interventions today provide ubiquitous examples of failed interventions is suggestive, and provides a directional constraint on what the prior probability of an effectiveness hypothesis ought to be (namely, lower than what currently seems to be the case).

In sum, many medical interventions have been rejected because they have later been found to be ineffective or harmful or both. This is most clear when considering interventions from history, but is also evident when considering contemporary medical interventions.

11.3 Medicine's Darkest Secret

For many medical interventions, the best evidence available today suggests that they are barely effective, if at all. The following examples illustrate the phenomenon.

---

[5] It was supposed to help with the filling of the lungs with fluid after heart failure. It also has no effect on mortality or rehospitalization rates; see (O'Connor et al., 2011). This large trial contradicted a set of smaller studies which suggested that nesiritide was helpful.

Medical interventions for which there is evidence of low effectiveness abound. As above, I focus on the widely used classes of interventions today. I take the title of this section from Marcia Angell, former editor of *The New England Journal of Medicine*, who claims that the paucity of new innovative drugs is the "darkest secret" of medicine (2004a).

Despite the widespread use of methylphenidate (Ritalin) for childhood attention deficit hyperactivity disorder (ADHD), there have been few studies of the long-term effectiveness of the drug for treatment of ADHD symptoms. The majority of trials of methylphenidate last only a handful of weeks. In the largest such trial, early findings suggested that methylphenidate is effective for childhood ADHD.[6] The active phase of this trial was fourteen months. Recall my discussion of trial duration in Chapter 9—trials are usually long enough to detect possible benefits of a drug but too short to detect many harms. Follow-up studies of the children in the trial, from three to eight years after the trial, show that there is no beneficial difference in any measured parameter between those children who had been on methylphenidate compared with those not on it. These parameters include core ADHD symptoms, grades earned in school, arrests, and psychiatric hospitalizations. However, those children on methylphenidate suffer from harms such as stunted growth: they are on average 2 cm shorter and 2.7 kg lighter than children on placebo. The only thing that predicted a child's symptoms and other social parameters six to eight years after the trial was the child's degree of symptoms at the start of the trial, regardless of treatment.[7] In other words, methylphenidate was useless at decreasing the symptoms of ADHD and its social ramifications.[8]

Treatment of depression with antidepressants is no better. In one of the most thorough meta-analyses on the effectiveness of antidepressants, Kirsch and his colleagues obtained data on all trials that had been submitted to the FDA—most importantly, this included data from unpublished trials. Recall from Chapter 8 that such trials employ the HAMD scale. This group had earlier found that the average decrease in HAMD score among all subjects was 1.8 after taking antidepressants (compared with placebo), which is clinically trivial given how the scale is constituted (remember: a HAMD score can go down by up to ten points if an intervention makes a subject sleep better and fidget less). Antidepressants have no effect for mildly depressed subjects and only a tiny effect for subjects with very severe depression. This is especially disappointing

---

[6] This was the MTA trial (Multimodal Treatment of Attention Deficit Hyperactivity Disorder Study), performed by the National Institute of Mental Health (NIMH).

[7] The references cited here are: (MTA, 1999), (Jensen et al., 2007), (Swanson et al., 2007), and (Molina et al., 2009). Although there is some evidence on the short-term effectiveness of methylphenidate, this literature is riddled with methodological problems. See the meta-analysis and discussion in (Schachter et al., 2001).

[8] The NIMH—the very organization that performed the MTA study—claimed on its website that methylphenidate is effective in helping to improve ADHD symptoms, academic performance, and other social problems, despite the evidence from their own study showing exactly the opposite (this page was online in 2013 but as of 2015 is no longer online).

given the problems with the scale and concerns about extrapolating from trials to uncontrolled clinical settings (see Chapter 8).[9]

The use of statins to lower cholesterol has increased enormously in recent years. For example, atorvastatin (Lipitor), sold by Pfizer, has been a best-selling drug for years, with sales over $125 billion. However, for people who have no history of heart disease, a review concluded that statins have "small and clinically hardly relevant" benefits, and another meta-analysis showed that, among patients receiving statins the overall risk difference for avoiding death was 1.2 percent—this implies that out of 100 people taking this drug for years, it will only benefit one of them.[10]

The recent approval of flibanserin for 'female sexual dysfunction' is another compelling example. Flibanserin was approved by the FDA for low sexual desire in women. A patient advocacy group called 'Even the Score' had accused the FDA of gender bias because the regulator had approved drugs for erectile dysfunction but had not yet approved a drug for female sexual desire (the patient advocates had been organized by a consultant to Sprout Pharmaceuticals, the developer of flibanserin). Critics claimed that low female sexual desire is not a condition constituted by a microphysiological abnormality, and thus not the kind of state for which a pharmaceutical is appropriate (to use the analysis from Chapter 2, it is not a genuine disease). But beyond the conceptual criticism, the empirical evidence for the alleged effectiveness of flibanserin was thin: on a six-point sexual desire scale, women on flibanserin increased 0.3 points compared with placebo, and had only one additional 'sexually satisfying event' per month.

Many prominent interventions are less effective than once thought. For instance, surgical reconstruction of torn anterior cruciate ligaments (ACL) is widely employed, but a recent study showed that this operation is no more effective than a non-surgical rehabilitation program. A prominent example of a medical intervention that appears to be much less effective than once thought based on the best evidence available today is oseltamivir (Tamiflu)—I return to this example below. A recent review found that cortisone shots for treating pain of overuse injuries (like tennis elbow) had worse outcomes than doing nothing or getting physical therapy, despite the fact that this has been a standard treatment for decades.[11]

One might charge me with cherry-picking examples of medical interventions that are barely effective. However, these examples are among the most widely used medical interventions today, and such examples are ubiquitous—it takes little effort to expand lists of such examples. In a summary opinion piece about effectiveness of drugs, for example, Spector (2010) notes that these entire classes of drugs are barely effective (despite the fact that some members of these classes have been approved by

---

[9] The references cited here are (Kirsch et al., 2002) and (Kirsch et al., 2008). The drugs in these meta-analyses included fluoxetine (Prozac, Sarafem), venlafaxine (Effexor), nefazodone (Serzone, Nefadar), and paroxetine (Paxil, Seroxat): all blockbuster antidepressants.

[10] The references cited here are (Vrecer et al., 2003) and (Baigent et al., 2005).

[11] The citations in this paragraph are: ACL: (Frobell, Roos, Roos, Ranstam, & Lohmander, 2010); oseltamivir: (Jefferson et al., 2012); cortisone shots: (Coombes et al., 2010).

the FDA): non-sedative antihistamines, anticholinergics, cholinesterase inhibitors, and leukotriene blockers.

A compelling illustration of the low effectiveness of many medical interventions can be represented by the outcome measure 'number needed to treat' (NNT). Recall my discussion of outcome measures in Chapter 8 (§8.3), where I gave the formal definition of this outcome measure. This is an intuitive measure: it states how many people would have to use an intervention in order to achieve one beneficial outcome. For example, if the NNT of an intervention is 50, then fifty people would need to use the intervention in order to achieve one positive outcome. In other words, only one of the fifty people who used the intervention would experience the beneficial outcome, while the other forty-nine would not. Thus the higher the NNT, the less effective the medical intervention is. A magic bullet (see Chapter 4) should have an NNT close to 1—nearly every person who uses a magic bullet, such as penicillin for certain bacterial infections, benefits from it (other than those who are allergic or resistant to penicillin, and unfortunately anti-biotic resistance is spreading). For a non-pharmaceutical example, consider defibril-lation during heart attack (often depicted in television), which has an NNT of 2.5 to avoid death. Unfortunately many contemporary medical interventions have extremely high NNTs, often over 100.

There is an evidence-based medicine group that collects evidence and reports the number needed to treat for various interventions on their website (thennt.com). Many of the interventions reviewed by this group show no benefit whatsoever—the NNT is infin-ity. Of those that do show some effectiveness, the NNT is often extremely high. Here are a few of the many examples of interventions reported by this group to have no benefit: statins after acute coronary syndrome to avoid stroke, heart attack, or death; statins to prevent death in patients with no known heart disease; prostate-specific antigen test, as a screen for prostate cancer, to prevent death; bisphosphonates for fracture prevention in post-menopausal women with no prior fractures; and routine health checks for reducing mortality and morbidity.[12] Here are a few of the many examples of interventions with high NNTs: anti-hypertensive drugs to prevent heart attacks (NNT = 100); statins to pre-vent heart attack in patients with no known heart disease (NNT = 104); magnesium sulfate to prevent seizure for women with preeclampsia (NNT = 90); long-acting beta-agonists with inhaled corticosteroids versus inhaled steroids alone, for adults with asthma, to prevent moderate asthma attack (NNT = 73). These pithy descriptions of low effectiveness do not include more subtle findings in the relevant reviews, including evi-dence of harms. These examples include some of the most widely used classes of drugs, such as long acting beta-agonists and statins, and widespread practices such as routine health checkups and cancer screening tests.

Another way to consider the ineffectiveness of medical intervention is to change the focus from the interventions themselves to the diseases for which we lack effective

---

[12] No screening program has been shown to decrease all-cause mortality, despite numerous empirical assessments of screening programs.

interventions. Most cancers are good examples. Since the so-called war on cancer began, the U.S. National Cancer Institute alone has spent over 100 billion dollars on cancer research, and yet the death rate for cancer dropped a mere 5 percent from 1950 to 2005 (Kolata, 2009). The death rate for most particular kinds of cancer has not changed (though for some particular cancers there has been remarkable progress). The only factor that seems to have an impact on decreasing cancer rates is cessation of smoking.

Not only is there widespread evidence of low effectiveness for medical interventions, but much evidence in medical research is discordant. For most interventions there are many studies that provide relevant evidence. The problem of discordance is that different studies often warrant contradictory conclusions: one study suggests that an intervention is effective while another suggests that it is ineffective. Discordance can be synchronic—at a given point in time the available evidence for a medical intervention is discordant—or discordance can be diachronic—at a given time the available evidence suggests that a medical intervention is effective but later evidence shows it to be ineffective.

Discordance is a pervasive feature of medical research, among all types of studies, including randomized trials and meta-analyses. Perhaps the most striking kind of discord is among meta-analyses. Different meta-analyses of the same primary evidence reach contradictory conclusions. In Chapter 6 I argue that discordance among meta-analyses is possible because the method is malleable. The effectiveness of antihypertensive drugs, methylphenidate for attention deficit hyperactivity disorder, spinal manipulation for back pain, and statins for primary prevention of cardiovascular disease—to name a few other examples—have all been variably estimated by discordant meta-analyses. Such malleability and the discordance it permits should be unsettling because meta-analysis is usually thought to be the final arbiter in medical research.

An example of discordance I return to below is that of oseltamivir (Tamiflu). When researchers attempted to review the available data to determine if claims about the effectiveness of oseltamivir were true, they found that "the public evidence base for this global public health drug is fragmented, inconsistent, and contradictory" (Doshi, 2009). Discordance can occur because later studies fail to replicate earlier studies. Ioannidis (2005a) showed how frequently this can occur: in a second-order study of forty-five blockbuster publications that purported to show that a medical intervention was effective, fourteen of these publications were later contradicted.

In sum, a dark secret of medicine is that the best evidence available today suggests that many new medical interventions are barely effective, and the available evidence is often discordant.

11.4 Master Argument

The many diverse problems with medical research articulated throughout this book can be unified into a single master argument, using a simple formal representation.

Call H a hypothesis that a medical intervention is effective, and call E the evidence about that intervention. One's confidence in the effectiveness of the intervention can be represented by a conditional probability—the probability that H is true (that the intervention is effective), given the evidence (E) that we have: P(H|E). By Bayes' Theorem, that is equivalent to the following (see Appendix 1):

$$P(E|H)P(H)/P(E)$$

This is a formal representation of the master argument for medical nihilism. One's confidence in the effectiveness of the medical intervention is directly proportional to P(H) and P(E|H) and inversely proportional to P(E). The arguments in this book conclude that on average the former two terms ought to be low and the latter term ought to be high, and thus P(H|E) ought to be low. Thus, one's confidence in medical interventions ought to be low.

Let us start by considering the prior probability, P(H). One central factor in determining the prior probability that an intervention is effective is the proportion of past interventions that have been effective. Since, as I argued in §11.2, so few past medical interventions have been effective, the prior probability that a medical intervention is effective ought to be low. Similarly, given my argument in Chapter 4 about the physiological basis for the paucity of magic bullets, for a typical intervention (prior to any evidence) we ought to have a low probability that the intervention will be effective. In other words, the dearth of magic bullets entails that a new medical intervention is very likely not a magic bullet nor does it even approximate the magic bullet ideal, and thus one's prior expectation that a new medical intervention is effective ought to be low. Thus, based on those two general reasons (pessimism from past failures and the dearth of magic bullets), for a typical medical intervention:

P(H) is low

The influence of ubiquitous evidence of low effectiveness on the 'likelihood,' P(E|H), is straightforward. If E is evidence that the intervention is relatively ineffective (say, the effect size is small), the probability of observing such evidence conditional on the truth of the hypothesis, P(E|H), is low. This is another way of saying that if the intervention were effective, we would expect to get evidence (from a trial, say) that the intervention is in fact effective. Small absolute effect sizes are widespread in medicine. Smaller effect sizes entail lower likelihood of effectiveness hypotheses.

The influence of discordant evidence on the likelihood is slightly less straightforward. Suppose some studies (call this set i) provide evidence that supports H; the likelihood of this evidence is $P(E_i|H)$. Suppose there are other studies that provide evidence that is discordant with evidence from i; these other studies provide evidence that the intervention is ineffective. Let the total set of available studies relevant to H be n ($i \in n$), and so the total evidence for H is E_n ($E_i \in E_n$). The likelihood of the total evidence is $P(E_n|H)$. Necessarily, $P(E_n|H) < P(E_i|H)$. The subset of evidence from i that confirms the hypothesis, E_i, is more likely to be observed if the hypothesis were true than would be the overall set of discordant evidence. In plainer terms, if the medical intervention were truly effective, then evidence that consistently shows that the medical intervention is effective would be more probable than would be discordant evidence. In short, discordance and small effect sizes entail that:

$$P(E|H) \text{ is low}$$

The influence of systematic bias on the prior probability of the evidence is straightforward. I described the formal representation of the influence of bias in §10.6. Given the extent of error, bias, and fraud in medical research, and given that these biases are systematically slanted toward the production of evidence that suggests that medical interventions are effective, when new evidence is published that suggests that a medical intervention is effective, we ought not be very surprised. Positive evidence for the effectiveness of an intervention is likely whether or not that intervention is indeed effective (at least, more likely than if such systematic bias were not present). In formal terms, the ubiquity of systematic bias in medical research entails that for typical evidence in medical research:

$$P(E) \text{ is high}$$

Since the two terms in the numerator of the master argument are low and the term in the denominator is high, it follows that the posterior probability is low:

$$P(H|E) \text{ is low}$$

The conclusion of the master argument is that one's confidence in the effectiveness of the medical intervention ought to be low. The conclusion of the master argument is medical nihilism.

In Chapter 1 I noted the various connotations that the term 'nihilism' can have: ontological, epistemological, justificatory, and emotional. The master argument for medical nihilism is based on ontological, epistemological, and justificatory reasons (and an emotional response might follow). Consider the argument about the dearth of magic bullets (see Chapter 4): this is an ontological reason that provides support to the master argument for medical nihilism. Or consider the argument about the malleability of research methods in medical science (see Part II): this is an epistemological reason that provides support to the master argument for medical nihilism. Or consider the concern about biases and fraud in medical science (see Chapter 10): this is a justificatory reason that provides support to the master argument for medical nihilism. I formulate the conclusion of the master argument in terms of the confidence that one ought to have in the effectiveness of medical interventions, and such confidence should be based on a range of ontological, epistemological, and justificatory reasons.

Some subtleties remain. First, although I am arguing that P(E|H) ought to be low and P(E) ought to be high, it remains the case that P(E|H) must be greater than P(E). That is because I stipulated that E is evidence that provides at least some confirmation to H, however minimal, which entails that P(E|H) > P(E). For the same reason, P(H|E) must be greater than P(H). Second, in the description of the master argument, H is a hypothesis about a medical intervention, with no specification about what specific type of intervention it is. However, we typically know that a medical intervention is a member of a particular class of interventions (say, a member of type T). The considerations that pertain to P(H)—namely, past failures and a dearth of magic bullets—may differ for members of T than for all drugs D. If the intervention in question were a new antibiotic that works similarly to penicillin, for example, then, contrary to the general arguments above, in this case P(H) ought to be fairly high; conversely, if the intervention in question were a new antidepressant, say, then P(H) ought to be relatively low.

The formal representation of the master argument is simply meant to unify the otherwise disparate arguments throughout this book. I do not pretend that the terms in the master argument can be precisely quantified, either generally or for a particular medical intervention. The conclusion of the master argument is necessarily vague. How low should one's confidence be in a medical intervention? There is no general, context-free answer to this question. However, the conclusion is in another sense very clear: given the proxy indicators of our confidence in medical interventions (see Chapter 1), the master argument for medical nihilism holds that this confidence should be much lower than is presently the case.

11.5 Objections

The thesis of medical nihilism is strong and invites some predictable counterarguments. In this section I articulate and dispel some of these objections.

11.5.1 The 'merely empirical' objection

Whether or not a medical intervention is effective, one might object, is simply an empirical matter. One need only perform a randomized trial on an intervention (or a meta-analysis of trials), this objection goes, and the results will tell us whether or not the intervention is effective. Obviously, hypotheses about the effectiveness of interventions are empirical, and thus there must be a significant empirical basis for evaluating such hypotheses.

But hypotheses about the effectiveness of medical interventions are not *merely* empirical. Effectiveness is a concept that itself depends on other difficult concepts such as health and disease, which themselves are partly empirical and partly normative (see Chapter 2); effectiveness hypotheses have particular scope and levels-of-effectiveness requirements (see Chapter 3); effectiveness has been articulated in a very particular model of interventions in the last century, and appealing to that very model diminishes the expectation of effectiveness for most interventions (see Chapter 4). Our best theory of scientific inference holds that to make an informed judgment about any hypothesis we must take into account more information than merely the empirical evidence available for that hypothesis (see Appendix 1). The evidence for medical interventions is manifold (see Chapter 5), and even the best evidence comes from malleable methods, typically bent toward overestimating effectiveness and underesti-mating harms of interventions (all chapters in Part II). The many pervasive biases in medical research exacerbate this (see Chapter 10). Finally, as noted in §11.3, the best empirical evidence available for many interventions supports the thesis of medical nihilism because such evidence is often discordant and demonstrates tiny effect sizes.

11.5.2 The 'medicine is awesome' objection

Another objection takes the form of a *reductio*: modern medicine is awesome, and so medical nihilism is wrong. One need only look at the greatest medical advancements, this objection goes, to be optimistic regarding medical interventions.

In Chapter 4 I articulated the magic bullet model of medical interventions and noted the fabulous advancements that exemplify this model. But upon describing this model it became clear that the vast majority of medical interventions introduced in the last sixty years do not come close to approximating this model. To be sure, there have been great medical discoveries, like the examples of insulin and penicillin that illus-trate the magic bullet model. But there have been hardly any magic bullets introduced in recent decades.

To illustrate, consider the following. *BMJ* invited its readers (mostly physicians) to nominate the most important medical discoveries since 1840, and it then compiled the top fifteen and asked its readers to vote for the best. The top medical breakthroughs were (in order of number of votes): sanitation, antibiotics, anesthesia, vaccines, discov-ery of DNA structure, germ theory, oral contraceptive pill, evidence-based medicine, medical imaging, computers, oral rehydration therapy, risk of smoking, immunology, chlorpromazine, and tissue culture. The list is an odd set of mixed categories. Several of these breakthroughs are methodological—they facilitate research or diagnosis and are only indirectly related to treatment, such as evidence-based medicine, medical imaging, and tissue culture. Several others are not themselves medical, though they have had an indirect impact on medicine—such as the discovery of DNA structure and computers. Anesthesia is not a therapeutic intervention, but rather is something done to facilitate treatment. Risk of smoking is a risk, not an intervention. Germ theory is a theory, not an intervention. Immunology is a scientific discipline, not an intervention. Oral contra-ceptive pill is an intervention to avoid pregnancy and not to treat a disease. Vaccines prevent the transmission of diseases rather than treat diseases. The only therapeutic interventions on this list are antibiotics and chlorpromazine.

This list is telling. Of the fifteen most important medical breakthroughs in the last 150 years, only two were considered to be therapeutic medical interventions. No statins, no SSRIs, no blood pressure drugs, no cancer drugs, no drugs for type 2 diabetes—in short, the most widely prescribed classes of interventions prescribed in recent years by physicians are not considered to be among the most important medical advances by physicians themselves.

Medications like streptomycin for tuberculosis have been so beneficial—the above objection goes—that it would be perverse to maintain a distrust of modern medicine.

The decline of infectious diseases since the nineteenth century in industrialized nations, one might claim, attests to the effectiveness of medical interventions. The problem with this objection is that medical interventions had little to do with the decline of infectious diseases. For instance, mortality due to tuberculosis had fallen by 75 percent *prior* to the introduction of streptomycin in the 1950s. Other infectious diseases such as pneumonia and diphtheria similarly declined long before pharmaceutical interventions were available for those diseases. The 'McKeown Thesis,' named after the physician and historian Thomas McKeown, was that the decline of mortality due to infectious diseases was due to an improved standard of living, which itself was a result of better economic conditions and specifically better nutrition—antibiotics had some role in this decline of mortality, but not as much as is often assumed.[13] The principal cause of the decline in infectious mortality was food, not pharmacy. Nevertheless, medical nihilism is not the absurd view that all medical interventions are ineffective, and many antibiotics are as close as our medicines get to being a magic bullet (see Chapter 4).

A recent study estimated the number of preventable hospital-caused deaths in the U.S. to be over 400,000 per year (James, 2013). Hospitals, then, are the third leading cause of death in the U.S., after heart disease and cancer. This should give pause to those raising the above objection, and indeed, is an independent consideration in favor of medical nihilism.

11.5.3 The regulation objection

One might object that a central premise of medical nihilism—the low effectiveness of many medical interventions—cannot be true, because (this objection goes) regulatory authorities like the FDA ensure that medical interventions available on the market are both safe and effective. This is a widely held view. Indeed, a study found that 39 percent of people in the U.S. agreed with the claim that the "FDA only approves prescription drugs that are extremely effective" (Schwartz & Woloshin, 2011). In Chapters 8 and 12 I argue that the FDA regulatory standard for medical interventions is too low, and amounts to underregulation. The arguments in those chapters are decisive rejoinders to this objection.

In short, unfortunately it is not the case that regulatory agencies ensure that only safe and effective medical interventions are publicly available.[14] For instance, the editor of the medical journal *The Lancet* accused the FDA of inadequately regulating alosetron (Lotronex), and that alosetron was reinstated despite evidence of serious harms due to the drug after confidential meetings with the manufacturer.[15]

---

[13] See (McKeown & Record, 1962), (McKeown, 1976a), and (McKeown, 1976b). I return to this idea in Chapter 12.

[14] See (McGoey & Jackson, 2009) and (Moynihan & Cassels, 2005).

[15] See (Horton, 2001). Approval by regulators does not imply that regulatory staff consider the intervention in question to be effective. Alosetron is such an example. It was approved by the FDA, despite the fact that senior scientists at the FDA thought that "the drug's meaningful benefits were on average non-existent

Here is another example. The manufacturer Wyeth recently reanalyzed its data from trials which tested its antidepressant venlafaxine (Effexor) in children; this reanalysis found an increased risk of hostility and suicidality among children taking venlafaxine. Wyeth voluntarily changed the drug's label to reflect this. In a report about suicidality caused by antidepressants published after this decision, the FDA had this to say: "FDA has not taken any regulatory action based on these findings for venlafaxine, since we view these as preliminary data" (Laughren, 2004). In other words, the company that makes venlafaxine and that has privileged access to evidence regarding the harms of this drug voluntarily changed its label based on evidence that the FDA described as preliminary.

Rosiglitazone is another case of regulators not providing adequate regulation. The FDA was aware of evidence that showed that rosiglitazone is harmful long prior to an independent meta-analysis that demonstrated such harms, yet the FDA did nothing with this evidence (see Chapter 9).[16]

Let me return to the example of oseltamivir. Many western governments stockpiled this drug out of concern for an influenza outbreak. A meta-analysis was published which amalgamated the results from ten manufacturer-funded trials—only two of which had been published—and based on this claimed that oseltamivir was effective at reducing influenza symptoms and associated complications (Kaiser et al., 2003). Subsequently, many regulators—in the United States, United Kingdom, Europe, and Australia—appealed to this meta-analysis as evidence for the effectiveness of oseltamivir. The U.S. Center for Disease Control and Prevention (CDC), for example, cited this meta-analysis as demonstrating that oseltamivir is effective. It was only when the Cochrane Collaboration initiated an update to their systematic review about oseltamivir that an independent physician posted a comment on the Cochrane website in which he noted that of the ten trials in the above meta-analysis, only two had been published in peer-reviewed journals. For years regulators and policy-makers relied on this misleading meta-analysis, which was used to help justify the massive expenditures for stockpiling oseltamivir. The Cochrane group later updated their review of oseltamivir with almost entirely negative conclusions about the drug's effectiveness.[17]

In short, regulators and other government agencies that provide guidance and policy in medicine do not ensure that available medical interventions are effective and safe.

---

or modest at best" (Moynihan & Cassels, 2005). Alosetron was withdrawn by the FDA because of its serious harms (but, oddly, later reinstated).

[16] Its inability to properly regulate is admitted by some senior FDA staff. For instance, one claimed that "we don't argue with drug companies; we listen to their distortions and omissions of evidence and we do nothing about it" (cited in Moynihan, 2002), and another claimed that "the FDA, as currently configured, is incapable of protecting America" (cited in Moynihan & Cassels, 2005). See also the worrying claims made by FDA staff quoted in §8.5.

[17] See (Jefferson et al., 2009) and (Jefferson et al., 2012). The CDC and the European equivalent were continuing to cite the misleading Kaiser meta-analysis in positive terms as late as 2011 (Doshi, Jones, & Jefferson, 2012).

11.5.4 The peer review objection

Some might object to medical nihilism on the grounds that medical science has checks and balances that mitigate misleading findings. Prominent among these checks and balances is peer review. Biased, misleading trials and meta-analyses that overestimate the effectiveness of medical interventions would not be published, goes this objection, because peer reviewers ensure that publications meet a high standard.

Unfortunately, peer review is insufficient to ensure that published research is reliable.[18] Another safeguard that one might expect to enhance the objectivity of medical science is the practice of replicating other scientists' results. Often, however, scientific results are not replicated, including even the most important of such findings. Ioannidis (2005a) provided an empirical demonstration of this: in a second-order study of forty-five publications (each of which purported to show that an intervention was effective), which were among the highest cited articles in the top medical journals over a thirteen-year period, eleven had never been replicated, and fourteen were later contradicted by attempts at replication. Thus, even among the absolutely best publications in medical science, over half the articles are either not replicated or are contradicted by later research. Richard Horton, editor at the *Lancet*, put the problem as follows: "we know that the system of peer review is biased, unjust, unaccountable, incomplete, easily fixed, often insulting, usually ignorant, occasionally foolish, and frequently wrong" (2000). Peer review and replication are insufficient to ensure that evidence suggesting the effectiveness of medical interventions is reliable.

11.5.5 Anti-science?

There is a prominent anti-science sentiment in contemporary society. A large proportion of Americans, for example, do not believe in evolutionary theory, and there is a widely held view that contradicts worrying findings from climate science. Such an anti-science sentiment can be very harmful.

Anti-science sentiments about medicine are widespread. For example, the anti-vaccine movement—prominently associated with a single publication that suggested that the measles-mumps-rubella vaccine can cause autism, which has been thoroughly discredited—has led many parents to not vaccinate their children, putting their own children and others at risk.

One might worry that the view presented in this book contributes to irrational anti-science sentiments. However, one would have to seriously misinterpret the message of this book to portray it this way. To make the master argument compelling, throughout this book I have appealed to high-quality science. The trouble with so much of medical research is not science per se, but poor reasoning based on low-quality science that suffers from many systematic biases, exacerbated by financial conflicts of interest.

---

[18] The iconic journal *Nature* put this point in an editorial as follows: "Scientists understand that peer review per se provides only a minimal assurance of quality, and that the public conception of peer review as a stamp of authentication is far from the truth" (Jennings, 2006).

To address some of these problems, I urge to make medical research more scientific, by increasing the standards of evidence in research (see Chapters 8 and 9), by carefully attending to the details of the science of pharmacology (see Chapter 4), and by employing our best theories of scientific inference (see Appendix 1). The problems articulated in this book that lend support to medical nihilism arise from a committed pro-science position.

11.5.6 Game-changers

One often reads of medical breakthroughs, blockbusters, or 'game-changing' interventions for particular diseases. In less cautious media such reports are very common. If game-changers are so common, then surely medical nihilism must be wrong?

We ought to be deeply suspicious of media reports of game-changers. Many reports of game-changers are drawn from conference proceedings or phase 1 or 2 trial results. Even worse, much of this discourse is based on animal studies prior to any evidence at all about effectiveness in humans. We have already seen how rampant publication bias is in phase 1 trials, and over 90 percent of medical interventions fail just at this stage alone. In any case phase 1 and 2 trials are much smaller than phase 3 trials, and so one should always be cautious when interpreting the results of such trials. Even most reports of phase 3 trials and meta-analyses should be interpreted with skepticism, given the arguments regarding choice of measuring instruments, subject selection, publication bias, and all the other methodological concerns presented throughout this book. Indeed, the master argument of medical nihilism concludes that game-changers will be rare, and unfortunately this has been borne out by the best empirical evidence since the introduction of the classical magic bullets. Talk of game-changers is almost always misleading.

11.6 Conclusion

This book has proposed numerous arguments that support medical nihilism, and the master argument brings these various arguments together in a unified structure. I have argued that our confidence that a medical intervention is effective ought to be low, even when presented with evidence for that intervention's effectiveness. How low? I do not think that there can be a precise or general answer. It is enough to say: lower, often much lower, than our confidence on average now appears to be. There is surprisingly little direct study of the confidence that physicians or patients or policy-makers have regarding the effectiveness of medical interventions. However, the confidence typically placed in medical interventions can be gauged by the resources dedicated to developing, marketing, and consuming such interventions. As noted in Chapter 1, by many indicators (number of drugs prescribed, amount of money spent on drugs, number of people on multiple drugs) our society is very confident in the effectiveness of medical interventions.

What explains the disparity between the confidence placed in medical interventions and the lower confidence that I have argued we ought to have? The ingenious techniques that companies use to market their products—paying celebrities to publicly praise their products, funding consumer advocacy groups, sponsoring medical conferences, influencing medical education, direct-to-consumer advertising—have been extensively discussed by others.[19] The promise of scientific breakthroughs partly explains this disparity—scientists seeking support for their research programs, and companies building hype for their products, often make bold predictions about the promise of the experimental interventions they are researching, and this can sound convincing when it is put in the language of genomics, proteomics, precision medicine, personalized medicine, and evidence-based medicine. Unwarranted optimism may be based in part on a history of a few successful magic bullets, such as penicillin and insulin—magic bullet thinking gets inappropriately adopted in premature proclamations of game-changing medical interventions, which media outlets promulgate. The sheer novelty of medical interventions seems to impress some.[20] For example, the class of drugs known as the second-generation antipsychotics has novelty built into its very name.[21] In short, successful interventions from early in the magic bullet era, powerful and clever marketing tactics, media hype, an expectation that new products must be better than old products—these may contribute to the gap between the confidence typically placed in medical interventions and that which is justified.

Medical nihilism is not the audacious view that there are no effective medical interventions. Indeed, the discussion in Chapter 4 articulated the magic bullet ideal and noted some interventions that approximate this ideal, such as penicillin and insulin. To this we could add other antibiotics, cisplatin for some forms of cancer, imatinib for chronic myelogenous leukemia, defibrillation for cardiac arrest, and many others. Effectiveness is a graded property: there are the best interventions, which approximate the magic bullet model, there are moderately effective interventions, there are those that are barely effective, and there are many that are not effective at all. To repeat: medical nihilism is not the thesis that there are no effective medical interventions. Please do not confuse this. Medical nihilism is, rather, the thesis that there are fewer effective medical interventions than most people assume and that our confidence in medical interventions ought to be low, or at least much lower than is now the case.

---

[19] Spending on pharmaceutical marketing increased from $11.3 billion in 1996 to $29.9 billion in 2005 (Donohue, Cevasco, & Rosenthal, 2007).

[20] In a defense of lithium against newer antipsychotic drugs, the psychiatrist Richard Friedman claimed that "New medical treatments are a bit like the proverbial new kid on the block: they have an allure that is hard to resist" (Friedman, 2009).

[21] Recently, though, systematic reviews have suggested that first-generation antipsychotics may in fact be superior. See, e.g., (Tyrer & Kendall, 2009), which has the salient title "The Spurious Advance of Antipsychotic Drug Therapy."

12

Conclusion

12.1 Gentle Medicine

The eminent physician Sir William Osler claimed that "The desire to take medicine is perhaps the greatest feature which distinguishes man from animals." Given the thesis argued for in this book, we ought to quell this desire. We should consume fewer medical interventions, physicians should prescribe fewer interventions, and policy-makers should approve fewer interventions. An extreme and implausible view would be absolute non-interventionism, which holds that one should never seek medical intervention. An equally implausible view would be to agree with most or all of the arguments presented in this book but to not change one's use of medical interventions. The arguments for medical nihilism motivate for patients a pronounced skepticism regarding medical interventions, and for physicians a humility regarding the therapeutic tools that they have at their disposal.

In Chapter 3 I discussed the notion of overtreatment—the view that many of us consume more medical interventions than we ought to. This has become a widespread view even within mainstream medicine. For example, the former editor of *BMJ*, Richard Smith, together with the health journalist Patrick Moynihan, titled an editorial with the provocative question "Too much medicine?" to which they answered "almost certainly" (2002). Medical nihilism supports the overtreatment thesis. If our confidence in medical interventions decreases, as I have argued it should, then we ought to consume less medicine. We are presently consuming too much medicine. We are overtreated.

Conservative approaches to treatment have been popular throughout history. Hippocratic texts from fifth century BC place emphasis on *vis medicatrix naturae* (the healing power of nature), and emphasize the priority of proper nourishment and exercise to achieve health. Numerous proclamations of non-interventionism were made—by artists, authors, scientists, and physicians—from the fifteenth to the nineteenth century. The twentieth century, in contrast, involved a radically aggressive approach to treatment. Even alternatives to mainstream medicine today involve forms of inter-vention (chiropractic treatment, herbal medicine, acupuncture). I co-opt a term sometimes used for this latter set of practices—*la médecine douce*, or gentle medi-cine. (To be clear, I am merely borrowing the term and not the associated principles or practices.) *La médecine douce* encourages a moderate form of therapeutic conservatism: it encourages medicine to be gentle.

Another quote from Osler illustrates the principle of gentle medicine: "One of the first duties of the physician is to educate the masses not to take medicine." Given the arguments for medical nihilism, treatment should be less aggressive, and more gentle. For the most commonly employed medical interventions—including many of the examples throughout this book, such as pharmaceuticals intended to lower cholesterol and blood pressure, or treat psychiatric conditions—the arguments for medical nihilism are most compelling. For these kinds of interventions, and for their associated diseases, we could use more of *la médecine douce*—fewer medical interventions and, perhaps, more lifestyle interventions (for example), and more *care*.

A movement called 'Choosing Wisely' aims to get doctors to stop using interventions that are not effective and safe (Malhotra et al., 2015). This movement has been adopted by many dozens of medical societies in many countries. This is the right direction for medicine. Gentle medicine is not the naïve view that people who are already on particular medications should come off them. We have a large volume of evidence on the effectiveness of initiating treatment with an intervention, tested in controlled experimental settings. But we have little reliable evidence about the effects of withdrawing treatments.

As an example of the sort of evidence we should have before widely implementing gentle medicine, consider the following study. Researchers applied a drug discontinuation program to a cohort of elderly patients who were taking an average of 7.7 medications. By applying standard treatment protocols and getting the consent of the patients and their physicians, the researchers discontinued 4.2 medications/patient for a total of 256 drugs. Of these, only six drugs (2 percent) were readministered because of a recurrence of symptoms. No harmful effects were attributed to the drug discontinuations, and 88 percent of the patients reported an improvement in health. Making medicine gentler would make us healthier.[1]

These suggestions are merely that—suggestive. How exactly should medicine be rendered more gentle? Recall from Chapter 1 that the nineteenth-century physician Jacob Bigelow—a medical nihilist who held that interventions were mostly ineffective at modulating diseases—claimed that one does not need medical interventions to alleviate suffering: "he who turns a pillow, or administers a seasonable draught of water to a patient, palliates his suffering" (1835). Simply caring for one who is suffering is valuable and helpful. Unfortunately, the conceptual space of gentle medicine has not been well explored—there has been an excessive focus on the magic bullet model of medical interventions (see Chapter 4), and relatively little research on the range of other, gentler options we could use to care for the diseased. To employ the analysis

---

[1] See (Jena, Prasad, Goldman, & Romley, 2015). Another study compared the mortality rate for patients with heart failure or cardiac arrest, during dates of cardiology conferences (when many of the senior cardiologists were out of the hospital at the conferences). This study found a lower mortality rate for patients with severe cardiac problems during cardiology meetings (Garfinkel & Mangin, 2010). The implication was that such patients do worse when treated by senior, well-known cardiologists. For commentary see (Emanuel, 2015).

from Chapter 2, medicine has construed CAUSAL BASIS OF DISEASE and NORMATIVE BASIS OF DISEASE narrowly, and has focused on the former. We have some (but not many) compelling examples of strategies consistent with gentle medicine, a few of which I note in §12.3, where I argue that one way in which medical research priorities should be modified involves devoting more resources to studying gentle interventions. In short, medicine should be more gentle, and learning how medicine could achieve this requires a realignment of the medical research agenda.

Gentle medicine is not the audacious proposal that physicians should not intervene at all. We have a few magic bullets in our arsenal and we should use them. Rather, gentle medicine is the more modest proposal that physicians should intervene less, perhaps much less, than is presently the case, and we should try to improve health with changes to our lives and to our societies.

12.2 Tweaking Methodological Details

Some of the arguments for medical nihilism that I have articulated are based on methodological shortcomings of medical research. In Chapter 6, for instance, I describe the malleability of meta-analysis, in Chapter 7 I note the variability of tools for assessing the quality of evidence in medical research, and in Chapters 8 and 9 I argue that many features of randomized trials generate systematic overestimations of the effectiveness of interventions and underestimations of their harms. A simple response to such problems would be to modify the research methods in order to resolve the noted shortcomings. This approach to medical nihilism I call 'detail-tweaking.'

Detail-tweaking is urged and developed by many in the evidence-based medicine community, and it is an important tactic. For example, in several chapters I have noted the ubiquity and nefarious epistemic influence of publication bias. Recently, editors of leading medical journals have demanded that trials be pre-registered prior to data collection, with the idea that those reviewing subsequent evidence can more properly assess an intervention if they can estimate the extent of publication bias, and this can be aided by a registry of trials.[2] Similarly, to mitigate p-hacking (see Chapter 8), methodological guidance is that the primary outcomes that are measured in a trial should be pre-specified and adhered to. To address the problem of poor measuring instruments, better instruments (more sensitive and specific) ought to be developed. In §12.4 I describe several other ways in which detail-tweaking could be helpful, in the context of regulation of drug approval, but since I have noted many ways in which methodological details can be tweaked throughout this book, I do not repeat the suggestions here.

---

[2] To address the concern about financial conflicts of interest (see Chapter 10), some journals and other organizations require that researchers disclose their financial ties. Such policies are insufficient to block the threat of conflicts of interest exacerbating biases in medical research—see (de Melo-Martín & Intemann, 2009).

Detail-tweaking would help many of the problems of medical research today. However, detail-tweaking would not mitigate the plausibility of medical nihilism. That is for several fundamental reasons. First, no amount of detail tweaking can change the complex basis of many diseases and the complex causal interactions of interventions on physiological systems (see Chapter 4). Second, as more detail-tweaking is implemented, estimates of the effectiveness of medical interventions will decrease. Shortcomings in medical research are systematically biased toward overestimating effectiveness, and thus if the shortcomings were mitigated then effectiveness estimates would decrease (see Chapters 8, 9, and 10). And if effectiveness estimates were to decrease, that would strengthen one of the central arguments for medical nihilism (see Chapter 11). Third, detail-tweaking cannot eliminate all biases from research. Methods can be rendered less malleable, and some biases can be mitigated (see Chapter 10), but no empirical methods are perfect, and when there are strong incentives for demonstrating effectiveness, biases are liable to be exploited to exaggerate claims of effectiveness. Fourth, detail-tweaking is irrelevant to some crucial problems in medical research, such as disease-mongering (see Chapter 3), the neglect of important strategies to improve health that do not follow the magic bullet model (see Chapter 4 and §12.3), and the proliferation of non-innovative me-too drugs (see §12.4).

In short, detail-tweaking will continue to be important in medical research, and as it is developed estimates of effectiveness of medical interventions will decrease, but detail-tweaking is insufficient to resolve the problems that motivate medical nihilism.

12.3 Rethinking Research Priorities

Medical nihilism holds that we ought to lower our confidence in the effectiveness of medical interventions. As a strategy to improve health, the magic bullet model has proven to be not nearly as helpful as many seem to suppose. Magic bullets have not lived up to the promise of the early successes like insulin and penicillin (see Chapter 4). Health is obviously an important goal and a fundamental aspect of human flourishing. How, then, should we be trying to maintain and improve health, given the arguments here for medical nihilism? What sorts of research projects should we be prioritizing?

One set of answers to these questions is available from work in social epidemiology and the social history of medicine. These disciplines are concerned with determining the causes of health, and the causes of inequalities of health. Within these disciplines there is near consensus that the most important causes of health and health disparities do not involve medical interventions, but rather involve broader features of society, such as access to clean drinking water and nutrition, and equitable distribution of societal resources. As mentioned earlier, Thomas McKeown argued that increases in longevity that began during the industrial revolution, and corresponding population growth, had little to do with medical interventions (such as antibiotics for tuberculosis) but rather were caused by better socio-economic conditions, and especially improved diet (1976a). Because people wrongly tend to think that interventions such as pharmaceuticals are significant causes of health, McKeown argued that people place far too much hope on the curative powers of medicine (1976b). Instead, we ought to be concerned with broader socio-economic conditions of health. The McKeown thesis, as it has been called, has attracted criticism. However, even the critics typically hold that significant health improvements such as increases in life expectancy have been due less to medical interventions and due more to broader social interventions such as increased access to clean drinking water.

A related thesis has been developed by the epidemiologist Michael Marmot (2004). Marmot argues that one's socio-economic status (position or rank in society) is a significant cause of one's health, even after controlling for other features of life that are associated with socio-economic status, such as income and lifestyle and access to quality healthcare. Marmot's theory is that people with higher social rank have a greater sense of control over their lives, and this increased feeling of autonomy positively influences health. A similar argument by Wilkinson (2006) holds that the mediator between inequality and poor health is stress. Given this causal relation between social rank and health mediated through stress or one's sense of self-control, we could work to modify structures of society in such a way that adds some balance to people's sense of social rank and that mitigates stress or increases people's sense of self-control.

In short, one implication of medical nihilism is that we could realign research priorities away from a focus on medical interventions such as pharmaceuticals and toward social interventions.

This suggestion should not be understood as the extreme view that there should be no research on potential new magic bullets. To the extent that resources are devoted to research on pharmaceuticals, such research should be on interventions that are likely to approach the magic bullet ideal (rather than, say me-too drugs with tiny effect sizes). Developing new antibiotics would be extremely valuable. With the development of bacterial resistance to antibiotics, we have fewer and fewer effective antibiotics at our disposal. Since infectious diseases can be lethal and spread quickly, and since antibiotics can be so effective, the development of antibiotic resistance is a major problem. Pharmaceutical companies have largely avoided research on new antibiotics in the last several decades, because patients tend to be on antibiotics for only a few days or weeks, unlike other kinds of drugs such as statins and antidepressants (which patients are on for many years), and because the threat of the development of resistance means that physicians have become cautious in prescribing antibiotics. Both of these reasons entail that antibiotics bring less profit to companies. Thus, the incentive for pharmaceutical companies to develop financially profitable drugs has created a perverse situation in that the most effective and needed kinds of pharmaceuticals—those most likely to approach the magic bullet ideal—are neglected.

Many commentators on biomedical research have noted what is called the 10/90 gap: only 10 percent of the world's medical research resources are devoted to health problems of 90 percent of the world's population (the world's poorest). Pogge (2005) notes that though the 10/90 gap might be an exaggeration, it represents an important and morally laden problem. To articulate this problem, Reiss and Kitcher (2009) apply the notion of 'well-ordered science'. They argue that for medical science to be well ordered, it must adhere to the 'fair-share' principle: the amount of resources devoted to a particular disease must be proportionate to the amount of suffering caused by that disease relative to other diseases. If disease d causes p percent of the global disease suffering, then p percent of medical research resources should be devoted to research on d. Medical research is far from well ordered in this sense.[3] Given that medical nihilism holds that the products of medical research are not very effective, especially those products that target the health problems of the wealthy, medical nihilism provides support to those who are concerned with the disproportionate amount of resources devoted to the health problems of the wealthy. The 10/90 gap together with medical nihilism suggests that research resource distribution ought to be modified (I address various proposals for such a modification in §12.5).

A proposal for modifying research priorities suggested by Ioannidis (2005b) in a groundbreaking article sounds at first glance odd. Ioannidis argues that most published research findings are false—a conclusion consistent with medical nihilism—based on considerations that are similar to the master argument, namely the prevalence of biased methods employed by researchers with conflicts of interest, hypotheses with low prior probabilities, and small effect sizes. One way to address this, suggests Ioannidis, is to study hypotheses with high prior probabilities. This would increase the posterior probabilities of hypotheses, and thus increase the chance that research findings are true. It sounds strange, because many hypotheses with high prior probabilities do not need more evidence. It would be wasteful, for instance, to test aspirin for relieving headaches, coffee for a morning perk, or parachutes for slowing a skydiving fall. On the other hand, testing hypotheses with low prior probabilities can also be wasteful and harmful. If we took into account the lessons of Part I of this book, then we would have little reason to think that there are magic bullets for many conditions, and thus a hypothesis that there is such a magic bullet ought to be assigned a low prior probability, and avoiding research on such a hypothesis could save vast amounts of research resources and minimize the introduction of ineffective and harmful interventions.

For example, based on the great complexity of the physical basis of mental states, chemical deficiency hypotheses of mood disorders should be assigned a low prior probability (see Chapter 2). This, together with the fact that interventions that modulate these chemical levels have great cascading complexity of physiological effects (see Chapter 4), entails that hypotheses about the effectiveness and safety of such interventions that target alleged chemical deficiencies of mood disorders should be assigned

---

[3] As Reiss and Kitcher note, for example, "malaria, pneumonia, diarrhea and tuberculosis together account for 21 percent of the global disease burden, but receive only 0.31 percent of all public and private funds devoted to health research."

a low prior probability. If we followed the suggestion of Ioannidis, we would not have pursued research into such interventions, and thus would have avoided the present situation in which interventions for mood disorders appear to be ineffective (see Chapter 11) and harmful (see Chapter 9).

Another way in which medical research priorities should be modified involves devoting more resources to discovering effective gentle interventions of the sort I mentioned in §12.1. Consider the following passage from a book written by an intensive care physician:

As the medical research community increasingly chased genes, proteins, and molecular pathways, it tended to ignore what it was like to be a human being confronted by illness. The laser-sharp focus on molecular medicine and the latest pharmaceutical therapeutics meant that research into the human side of life-threatening illness received short shrift.   (Brown, 2016)

The focus on magic bullets ("genes, proteins, and molecular pathways") has on the whole been a disappointment (see Chapter 4). Gentle medicine is concerned with attending to what it is like to be a person confronting disease. Other sorts of gentle interventions that should be more fully explored include lifestyle interventions. For example, in a systematic examination of the benefits of exercise compared to drug interventions, Naci and Ioannidis (2013) found that there was no detectable difference in mortality rates between exercise and drug interventions for the secondary prevention of coronary heart disease and prediabetes, and exercise was more effective than drugs at reducing mortality among patients with strokes. We have very little high-quality evidence that investigates these sorts of strategies for improving health. We should have more.

12.4 Regulation and Inductive Risk

A reasonable response to medical nihilism is to demand enhanced scrutiny of medical interventions by regulators. In Chapter 8 (§8.5) I described the FDA approval process and argued that it suffered from many measurement problems. Here I articulate this regulatory standard in terms of an 'inductive risk calculus' and suggest ways in which this inductive risk calculus can be improved.

Some critics argue that the FDA overregulates the introduction of new pharmaceuticals. These critics hold that the epistemic standards required for new drug approval are cumbersome, disincentivize research into new pharmaceuticals, and raise the prices of drugs.[4] Other critics argue that the FDA underregulates the introduction of new pharmaceuticals. These critics hold that the epistemic standards required for new drug approval are too low and allow drugs that are ineffective or unsafe to be approved. Such criticisms have been voiced by academic scientists (such as Steve Nissen, who performed

---

[4] Such criticisms tend to come from free-market economists or institutions. See, for example, (Friedman & Friedman, 1990) and (Becker, 2002).

the 2007 meta-analysis on rosiglitazone, discussed in several chapters in this book), scientific organizations (such as the U.S. Institute of Medicine), and even by staff within the FDA (such as the epidemiologist David Graham).

Just as in the prominent discussion of inductive risk presented by Rudner (1953) and extended by Douglas (2000) and others, non-epistemic values play a role in setting epistemic standards in drug regulation.[5] When assessing the effectiveness and safety of a pharmaceutical, one is liable to make a false inference based on the available evidence—one faces inductive risk. At least some experimental pharmaceuticals are effective (though many are not), and few experimental pharmaceuticals are completely safe since most cause at least some unintended harmful effects. Regulators must make a judgment about the relative effectiveness-harm profile of an experimental pharmaceutical, based on whatever evidence they have. To do this, regulators must make an inference, and there are two fundamental errors they can make: they can infer that a drug has a favorable effectiveness-harm profile when it in fact does not, or they can infer that a drug does not have a favorable effectiveness-harm profile when it in fact does. The former kind of error leads to unwarranted drug approvals and can harm patients by allowing relatively ineffective and unsafe drugs to be available. The latter kind of error can lead to unwarranted drug rejections and can harm patients by prohibiting relatively effective drugs from being available and can harm the financial interests of the manufacturer of the drug.

To avoid these two kinds of errors regulators employ numerous tactics. Many of these tactics trade off against each other, in that employing a tactic to decrease the probability of committing one of the error types increases the probability of committing the other error type. For example, demanding more positive trials for drug approval decreases the probability of unwarranted drug approvals but increases the probability of unwarranted drug rejections. To take an extreme case, a tactic to guarantee that regulators never commit the error of unwarranted drug rejections is to approve all new drug applications, thereby greatly increasing the probability of unwarranted drug approvals; and vice versa, a tactic to guarantee that regulators never commit the error of unwarranted drug approvals is to reject all new drug applications, thereby greatly increasing the probability of unwarranted drug rejections. We can conceptualize a scale of inductive risk: on one end of the scale is certainty that the error of unwarranted approvals is avoided (and thus a high probability that the error of unwarranted drug rejections is committed) and on the other end of the scale is certainty that the error of unwarranted drug rejections is avoided (and thus a high probability that the error of unwarranted drug approvals is committed).

Regulators must determine where their policies stand on this scale of inductive risk. This is an *inductive risk calculus*. Non-epistemic values influence this inductive risk calculus. The criticisms of FDA overregulation and underregulation can be understood

---

[5] For a sampling of the recent literature on values in science, see (Steel, 2010), (Elliott, 2011), (Steele, 2012), and (Wilholt, 2012).

in terms of this calculus: some critics hold that the FDA's inductive risk calculus places its regulatory stance too far toward the extreme of never committing the error of unwarranted drug approvals (overregulation), whereas other critics hold that the FDA's inductive risk calculus places its regulatory stance too far toward the other extreme of never committing the error of unwarranted drug rejections (underregulation).

With the benefit of resources from earlier chapters, it is simple to articulate numerous problems with the regulatory standard. The standard is loose regarding what parameters must be measured in trials (see Chapter 3) and what instruments can be used to make such measurements (see Chapter 8), the standard does not rely on a good statistical measure of effectiveness (see Chapter 8), the standard neglects the fact that research leading up to a new drug application does not adequately assess the harm profile of drugs (see Chapter 9), the standard can be met by trials which suffer from numerous biases (see Chapter 10), and the standard ignores the ubiquity of publication bias (see Chapters 5, 8, and 10). These problems lend support to those who challenge the FDA with underregulation. Below I suggest some ways in which the inductive risk calculus can be retuned to address some of these problems.

Consider Kitcher's notion of well-ordered certification in the context of inductive risk (2011): ideal deliberators pondering an inductive risk calculus—taking into account the relevant non-epistemic values of patients and manufacturers and society at large—would demand a balanced stance on an inductive risk calculus for drug approval. The fundamental way in which the FDA's inductive risk calculus could achieve more balance is to require more and better evidence regarding the effectiveness and harms of new pharmaceuticals. There are some straightforward tactics to achieve this.

To address the problem of p-hacking (see §8.5), more appropriate quantitative measures of effectiveness should be employed as standards for drug approval, such as the risk difference measure (see Chapter 8). The risk difference should be large enough that a typical patient with the disease in question could expect to receive substantial benefit from the intervention on an important patient-level parameter which is pertinent to the disease being treated (more precisely, the risk difference should be large enough such that approval and use of the medical intervention could be expected to maximize utility). Moreover, trial designs and analytic plans, including the choice of primary outcomes to be measured, should be made public in advance of the trial, and departures from the design or analytic plan should mitigate the assessment of the quality of the evidence by regulators.

Before a new drug application is approved, trials should show that the drug is effective and relatively safe in a broad range of subjects that represents the diversity of typical patients who will eventually use the drug in real-world settings. Trials should be designed to rigorously examine the harm-profile of experimental drugs, and should employ measurement instruments that provide faithful representations of the disease in question.

To address publication bias, all trial data should be publicly available, and trial registration should be necessary, and enforced by regulators. The inductive risk calculus of regulators should incorporate all evidence from all trials. To mitigate the concern about financial conflicts of interest influencing subtle aspects of trial design in a biased manner, regulators should require evidence from trials performed by organizations that are entirely independent of the manufacturer in question (such as a university or another government agency)—I discuss this in §12.5.

There are structural problems with the way the FDA is organized and funded and how it relates to industry.[6] Much of the funding of CDER comes from user fees paid by industry to have their new drug applications evaluated, and critics claim that since these user fees pay the salaries of drug reviewers, reviewers are beholden to the sponsors of new drug applications. Moreover, the FDA relies on advisory committees, which are composed of internal staff and external scientific consultants, and these committees often have significant conflicts of interest.[7] Finally, critics note that CDER contains both the office that approves new drugs and the office that tracks the harms of drugs that have been approved, which creates an institutional conflict of interest, because once CDER has approved a drug there is a strong disincentive to admit that it made a mistake by paying heed to the office that tracks the harms of approved drugs. Thus, a general way to improve regulation in the U.S. would be to modify the institutional structure of the FDA.

A predictable response to the claim that regulation of medical interventions should be enhanced is that more regulation disincentivizes research and innovation, thereby decreasing the number of valuable interventions that become available or delaying when they become available.[8] However, the thesis of this book should give pause to this objection to regulation. Medical nihilism entails that there are not many effective interventions anyway, regardless of regulation. Moreover, many new interventions are merely me-too drugs—drugs that are very similar to pre-existing drugs and that often have trifling effectiveness. Finally, there is reason to think that increasing the evidential standards for drug approval could improve innovation: as critics of the pharmaceutical industry note, profit can be had by effective marketing rather than effective drugs, but if regulators increased their epistemic standards, the profit incentive would remain,

---

[6] The FDA epidemiologist David Graham claims that the "FDA is inherently biased in favor of the pharmaceutical industry. It views industry as its client, whose interests it must represent and advance. It views its primary mission as approving as many drugs it can, regardless of whether the drugs are safe or needed" (2005).

[7] Resnik (2007) and Krimsky (2003) discuss an investigation that examined meetings by FDA advisory panels: in the vast majority of meetings there was at least one panel member with a financial conflict of interest, and over half the panel members in most of the meetings had financial interests that were "directly related to the topic of the meeting" (Resnik, 2007). See Chapter 10.

[8] As the economist Gary Becker puts it "New medicines are a major force behind the rapid advances in both life expectancy and the quality of life that have come during the past 50 years" (2002); increasing the epistemic standards for drug approval, according to such views, amounts to hindering the development of new drugs, and thus amounts to hindering our health. Even the present commissioner of the FDA, Robert Califf, seems to hold a view like this—in a recent presentation Califf included a slide that claimed that regulation is a barrier to innovation.

so enhanced regulation could spur companies to develop more effective drugs to meet the higher standard.

A related counterargument is that drug development is already very costly, and increasing the epistemic standard for drug approval will increase this cost. This cost would be passed on to patients, and since many drugs are already expensive—goes this objection—more regulation will make the expense of drugs even more burdensome.[9] This counterargument is unconvincing for a number of reasons. It is not solely the cost of drugs that matters to patients or to payers. Payers and consumers care about a more complicated property of drugs than cost, namely, the benefit accrued to the patient due to the effectiveness of a drug relative to the cost of the drug and the harms caused by the drug. In order to assess this more complex property, we must have more and better evidence regarding the effectiveness and harmfulness of drugs. Moreover, many of the proposals suggested above for modulating the regulatory inductive risk calculus are simple suggestions that would not add significant costs to development. Further, the concern about cost to consumers is misguided, since the bulk of the expense of new drugs is a result of the temporary monopoly granted to manufacturers of new medical interventions enabled by the patent system (see §12.5).

An interesting proposal to address some of the problems with the imbalanced inductive risk calculus of regulators is what Biddle calls "adversarial proceedings for the evaluation of pharmaceuticals" (2013). Based on Arthur Kantrowitz's notion of a science court, this would involve two groups of interlocutors debating the merits of a drug, where one group would be appointed by the sponsor of a drug and the other group would be composed of independent scientists, consumer advocates, and prior critics of the drug. The proceedings would be run by a panel of judges, who would come from a variety of scientific disciplines and would be independent of the drug's sponsor.[10] The idea would require many details of implementation to be worked out, but it could alleviate some of the problems associated with the imbalanced regulatory inductive risk calculus, though in the following section I suggest going one step further by taking all evaluation of medical interventions out of the purview of the manufacturers of those interventions.

12.5 Revolutionizing Medical Research

One might be pessimistic about the potential for the above suggestions—tweaking methodological details (see §12.2), rethinking research priorities (see §12.3), and

---

[9] Some estimates hold that new drugs, on average, cost over $500 million to get FDA approval (Resnik, 2007), but others argue that this estimate is grossly inflated because the estimate includes business activity, which is better thought of as marketing rather than research (Angell, 2004b).

[10] To Biddle's proposal I would add that philosophers of science—trained in scientific reasoning and nuances of medical research—would be a valuable addition to such panels. Biddle's proposal can be motivated by work in feminist epistemology, which argues that epistemic standards can be enhanced by including diverse perspectives in scientific evaluation; see (Longino, 1990) and (Wylie, 1992); de Melo-Martin and Intemann (2011) apply this principle to the context of medical research.

realigning regulation (see §12.4)—to adequately address the problems that motivate medical nihilism. If so, one might think that more is required. Some have suggested large-scale changes to the social and legal context of medical research. For example, Brown (2008) argues for the socialization of medical research and the elimination of patents for the discoveries of medical research. Brown gives two arguments for this. His first argument is based on the idea that medical research is foundational for clinical treatment, and so political considerations in favor of socializing the delivery of medical treatment are also considerations in favor of socializing medical research. Since most people are convinced by the former, they also should be convinced by the latter. His second argument is empirical: patent-protected medical research has not been very successful, whereas other areas of science such as physics and astronomy, and even great discoveries in medicine, such as Salk's polio vaccine, have thrived in the absence of financial incentives generated by intellectual property laws.

A predictable objection to Brown's proposal for eliminating patents is that such a policy would eliminate a strong incentive to perform medical research, namely the motive of profit. If one takes this motive away, goes this objection, then there will be fewer new drugs for patients. However, this objection is unconvincing for at least three reasons. First, the central thesis of this book is that there is a dearth of effective interventions anyway, and thus abandoning intellectual property protection for medical discoveries will not have much effect on the pipeline of effective new inter-ventions. Second, many new drugs introduced in recent years—spurred by a profit motive—are 'me-too' drugs, which are not significant advances in the treatment of diseases but rather are products which are very similar to existing drugs on the market. Third, the objection relies on a classic justification for intellectual property laws, namely the utilitarian justification that intellectual property laws promote innovation. However, this is an empirical premise, which, as Brown (2008) and Angell (2004a) argue, is unjustified in the case of medicine and perhaps science more generally (patents can in fact hamper research, as Biddle (2014) and others argue and as the case of me-too drugs suggests).

Great scientific advances—such as Galileo's discovery of the moons of Jupiter, Darwin's discovery of evolution by natural selection, and Einstein's theory of general relativity—were not spurred by a profit motive. This motive for discovery without financial incentive has been important in medicine. The most important discoveries in medicine, such as the germ theory of disease, were not motivated by profit, and most of the magic bullets (see Chapter 4), and other great medical discoveries such as the X-ray, were not protected by patent. Banting and Best, discoverers of insulin (one of the magic bullets from Chapter 4), did not seek a patent for their discovery—having witnessed its capacity to save lives, they believed that insulin should be available to anyone who needed it. Alexander Fleming also gave up his patent for penicillin to ensure its widespread availability.[11] Jonas Salk, developer of the polio vaccine—another

---

[11] See (Gabriel, 2014) for a detailed historical study of patents in medicine at the end of the nineteenth century and beginning of the twentieth century.

of the greatest medical discoveries of the twentieth century—was asked who owned the patent for the vaccine, to which he responded "There is no patent. Could you patent the sun?"

Reiss and Kitcher (2009) agree with Brown that intellectual property laws governing medical research should be modified. Rather than eliminating patent protection, however, their proposal is to gradually decrease the duration of patent protection, while examining the effects of such a decrease on pace of innovation. (In the United States today new medical interventions are protected for twenty years.) To address the problem of the 10/90 gap noted in §12.5, Reiss and Kitcher propose a global health institute, which would focus on the neglected diseases that tend to cause suffering to the world's poorest. This is in contrast with Pogge (2005), who proposes a scheme to reward industrial pharmaceutical research to the extent that the products of such research mitigate global disease burden. Reiss and Kitcher also suggest more modest proposals, such as eliminating user fees paid to the FDA by sponsors of new drug applications. They rightly claim that the U.S. law governing patents that requires interventions to be both 'genuinely novel' and 'useful' should be enforced—medical nihilism holds that many interventions are not nearly as useful as many seem to think, and properly enforcing intellectual property laws would eliminate patent protection of ineffective interventions—this could further spur the development of truly useful interventions.

Trials that are funded by centralized government agencies tend to be better quality than privately funded trials. Consider the MTA trial on methylphenidate for ADHD, discussed in Chapter 11. It was of a much longer duration than most industry-funded trials, and there was diligent follow-up of patients. Large, publicly funded trials are better able to mitigate the threat of biases, and since those biases are systematically skewed toward overestimating effectiveness and underestimating harms of medical interventions, publicly funded trials typically demonstrate smaller effect sizes (as the MTA trial did). In Chapter 10 I described a second-order study of trials funded by the National Heart, Lung, and Blood Institute, which since 2000 have had an average risk difference of zero—the interventions tested in these trials provided no benefit to the subjects. Publicly funded trials are more truth-conducive. One could disagree with Brown that we should eliminate intellectual property for medical interventions, or one might agree with Brown but cynically maintain that such a change will not happen anytime soon. Nevertheless, one could agree that on epistemic grounds, medical interventions should not be tested by their manufacturers—there is simply too much conflict of interest, which exacerbates the many biases present in medical research. Instead, interventions should be tested by independent academic researchers or government agencies.

12.6 The Art of Medicine and the Love of Humanity

I close this book with a quote attributed to Hippocrates, the symbolic parent of western medicine: "Wherever the art of medicine is loved, there also is love of humanity." This book has been written in that spirit. My hope is that a love of humanity can motivate us to improve the art of medicine—in all of its manifestations, including clinical practice, scientific research, and regulation—and conversely, rethinking the art of medicine could contribute to improving the condition of humanity. Medical nihilism suggests that improving the art of medicine will involve less focus on the development of new medical interventions in the model of magic bullets, and more focus on the modification of social structures and tactics to make medicine gentler, and ultimately more effective.

Medical nihilism holds that medical interventions as conceived of in the last century—as magic bullets—are too often not effective for improving and sustaining our health, or at least not as effective as we tend to think. Medicine should be more gentle. A hint is offered by another claim attributed to Hippocrates: "walking is our best medicine." For those ailments that affect so many in wealthy parts of world, which have been the focus of the medical nihilism defended in this book—such as type 2 diabetes, heart disease, and depression—the wisdom of Hippocrates remains relevant. For ailments that affect so many in more impoverished parts of the world, the lessons of social historians of medicine and social epidemiologists apply: the most effective interventions for improving health in such contexts are things like mosquito nets, clean drinking water, better nutrition, and greater socio-economic equality. We would be better off shifting our attention away from the magic bullet model of medical interventions as developed in the last century and toward ways of restructuring society that truly benefit the health of humanity.

Appendix 1: Bayes' Theorem and Screening

Here I introduce Bayes' Theorem and illustrate it with an example of a screening test. This example also highlights a concern I discuss in §3.5: namely, that screening, even with very reliable screening tests, can lead to many false positive diagnoses.

Consider a screening test for disease D which is accurate 99 percent of the time: if a person has D the test indicates the presence of D 99 percent of the time and if a person does not have D the test indicates the absence of D 99 percent of the time. Suppose D occurs in 1 person in 5000. If a person is screened with the test and the test indicates that the person has D, what is the probability that the person has D?

I will return to this question in a moment. First I introduce an important tool that I rely on throughout this book. This tool is used by philosophers of science for many purposes, including modeling scientific reasoning and defining important scientific concepts such as causation and explanation, and it can also be used to answer questions like the one above. The tool is called a conditional probability. Conditional probabilities take the form of a statement "the probability of A, given B"—the probability that it is raining outside, given that I am now in England, for example. It is helpful to have an efficient way to express conditional probabilities, so I write P(A|B) to represent the probability of A given B. Probabilities must satisfy certain axioms, and there is a convenient theorem, called Bayes' Theorem, which allows one to express conditional probabilities in useful ways. Here is the theorem:

$$P(A|B) = P(B|A)P(A) / P(B)$$

One can find easy proofs of Bayes' Theorem in probability textbooks or online.[1] Another useful theorem (the principle of total probability, which I describe in slightly more detail in Chapter 10) gives another version of Bayes' Theorem. The tilde symbol '~' represents negation: thus P(~X) means "the probability of not X." Here is the second version of Bayes' Theorem:

$$P(A|B) = P(B|A)P(A) / \left[P(B|A)P(A) + P(B|\sim A)P(\sim A) \right]$$

This is a very general and powerful theorem. I put the two versions of it to various uses throughout this book.

Note that the question about the screening test that I started with is a conditional probability: what is the probability that the person has the disease, given that the test indicates that the person has the disease? Let T be the situation in which the test indicates that a person has D. Put in formal terms, our question is to determine the probability that D, given T, or P(D|T). Since this is a conditional probability, we can apply Bayes' Theorem:

$$P(D|T) = P(T|D)P(D) / \left[P(T|D)P(D) + P(T|\sim D)P(\sim D) \right]$$

---

[1] For applications of Bayes' Theorem to philosophy of science, see (Howson & Urbach, 1989), (Earman, 1992), and (Bovens & Hartmann, 2004).

We were given enough information in the statement of the question to determine this quantity. We know that P(T|D), the probability that the test would indicate that a person has D given that they in fact have D, is 0.99. We know that P(D) is 1 in 5000, or 0.0002. We know that P(T|~D) is 0.01. And we know that P(~D) is 4999 in 5000, or 0.9998. So,

$$P(D|T) = 0.99 \times 0.0002 / [0.99 \times 0.0002 + 0.01 \times 0.9998]$$
$$= 0.02$$

That is, if a person is screened with the test and the test indicates that the person has D, there is only a 2 percent chance that this person actually has the disease.

Appendix 2: Measurement Scales

The typology of scales used in the argument of §5.6 is standard (Suppes & Zinnes, 1962). Consider the following examples of four different kinds of comparative measurements about two restaurants, Kiribati Kuisine and Tahitian Treats:

(i) Beth claims that the food at Kiribati Kuisine is better than at Tahitian Treats
(ii) The temperature inside Kiribati Kuisine is 20°C while the temperature inside Tahitian Treats is 22°C
(iii) Kiribati Kuisine has been in business five years longer than Tahitian Treats
(iv) Kiribati Kuisine has fewer items on its menu than does Tahitian Treats

The scales of these measurements are, respectively, ordinal (i), cardinal (ii), ratio (iii), and absolute (iv). Scale types can be defined by the sorts of transformations to the measured quantities that preserve the information in the measurement. Any positive transformation to a measure of Beth's food tastes will preserve the information in (i). Only a positive linear transformation will preserve the information in (ii)—for instance, we could switch to the Kelvin scale. Similarly, only a positive linear transformation will preserve the information in (iii)—for instance, we could switch to a scale of weeks instead of years—and there is a natural zero point: the date of business inception. Only an identity transformation will preserve the information in (iv): the number of items on each menu. From a measure on a cardinal scale—take the one in (ii), for example—we can infer a measure on an ordinal scale—in this example, that it is colder inside Kiribati Kuisine than it is inside Tahitian Treats. But such inferences cannot be reversed: we cannot take information from an ordinal scale, like that in (i), and infer information on a cardinal scale, like that in (ii). The amount of information in these scales increases from (i) to (iv).

Appendix 3: Epistemic Proof of Superiority of RD over RR

I use the definitions of RR and RD represented by conditional probabilities (Chapter 8). By applying Bayes' Theorem, RR is equivalent to:

$$RR = \left[P(E|Y)P(Y)/P(E)\right] / \left[P(C|Y)P(Y)/P(C)\right]$$
$$= \left[P(E|Y)/P(E)\right] / \left[P(C|Y)/P(C)\right]$$

The baseline probability of having outcome Y, P(Y), has fallen out of the equation. Thus RR is not sensitive to P(Y).

In contrast, consider RD. By applying Bayes' Theorem, RD is equivalent to:

$$RD = \left[P(E|Y)P(Y)/P(E)\right] - \left[P(C|Y)P(Y)/P(C)\right]$$
$$= P(Y)\left[\left[P(E|Y)/P(E)\right] - \left[P(C|Y)/P(C)\right]\right]$$

The leftmost multiplicand just is the prior probability of Y. Thus RD is sensitive to P(Y). The rightmost multiplicand is a representation of the extent to which consuming the intervention changes the probability of Y. One can see this by applying Bayes' Theorem once again, to the rightmost multiplicand:

$$RD = P(Y)\left[\left[P(Y|E)/P(Y)\right] - \left[P(Y|C)/P(Y)\right]\right]$$

The terms in the rightmost multiplicand are intuitive representations of the difference-making capacity of the experimental intervention—P(Y|E)/P(Y)—and control intervention—P(Y|C)/P(Y)—respectively.

Appendix 4: Decision-Theoretic Proof of Superiority of RD over RR

I use the definitions of RR and RD represented by conditional probabilities (Chapter 8). Let A mean that a patient consumes treatment A; let B mean that the patient consumes treatment B (this could be a competitor intervention, or placebo, or nothing at all); let a be the cost of consuming A (where cost is construed broadly, to include all harmful effects of A); let b be the cost of consuming B (again construed broadly); let Y mean that the outcome of interest occurs; finally, let the utility of Y be u and the utility of ~Y be u'. The expected utility of consuming A is EU[A] and the expected utility of consuming B is EU[B].

An outcome measure is EU-sufficient if and only if the outcome measure is sufficient to compare EU[A] and EU[B], for given a, b, u, and u'. An outcome measure is EU-insufficient if and only if it is not EU-sufficient (that is, if and only if the outcome measure is insufficient to compare EU[A] and EU[B], for given a, b, u, and u'). If an outcome measure is EU-sufficient then there is a strong pro tanto reason for requiring its use in measuring the effectiveness of medical interventions, and conversely, if an outcome measure is EU-insufficient then there is a strong pro tanto reason against its use in measuring the effectiveness of medical interventions. I now prove that RD is EU-sufficient and RR is EU-insufficient.

RD is EU-sufficient

$$EU[A] = P(Y|A)u + P(\sim Y|A)u' - a$$
$$= P(Y|A)u + [1 - P(Y|A)]u' - a$$
$$= P(Y|A)(u - u') + u' - a$$

$$EU[B] = P(Y|B)u + [1 - P(Y|B)]u' - b$$
$$= P(Y|B)(u - u') + u' - b$$

The expected utility of consuming A rather than consuming B is:

$$EU[A] - EU[B] = P(Y|A)(u - u') + u' - a - [P(Y|B)(u - u') + u' - b]$$
$$= [P(Y|A) - P(Y|B)](u - u') - (a - b)$$

Note that RD appears as the leftmost multiplicand in this term. Since the only other terms are the given costs and utilities, this shows that RD is EU-sufficient. I will put this in slightly different terms. To maximize expected utility a patient should consume A rather than B if and only if the expected utility of consuming A is greater than that of consuming B. The corresponding decision rule is:

> (*) For any u, u', a, and b (without loss of generality: a > b and u > u'), consume A rather than B iff EU[A] > EU[B]

Note that

$$EU[A] > EU[B] \text{ iff } \left[P(Y|A)(u-u') + u'-a \right] > \left[P(Y|B)(u-u') + u'-b \right],$$

which is equivalent to

$$EU[A] > EU[B] \text{ iff } \left[P(Y|A)(u-u')-a \right] > \left[P(Y|B)(u-u')-b \right]$$

which is equivalent to

$$EU[A] > EU[B] \text{ iff } \left[P(Y|A)-P(Y|B) \right] > (a-b)/(u-u')$$

Thus (*) holds that one should consume A iff $[P(Y|A) - P(Y|B)] > (a - b)/(u - u')$. RD appears on the left side of this inequality, and the right side of the inequality is determined by a, b, u, and u'. If given a, b, u, and u', then RD is sufficient to determine if the inequality in (*) holds. Thus RD is EU-sufficient.

RR is EU-insufficient
Above I showed that

$$EU[A] > EU[B] \text{ iff } \left[P(Y|A)-P(Y|B) \right] > (a-b)/(u-u')$$

Note that

$$P(Y|A)-P(Y|B) = P(Y|B)\left[P(Y|A)/P(Y|B)-1 \right]$$

which is equivalent to

$$P(Y|A)-P(Y|B) = P(Y|B)(RR-1)$$

and so

$$EU[A] > EU[B] \text{ iff } P(Y|B)(RR-1) > (a-b)/(u-u')$$

which is equivalent to

$$EU[A] > EU[B] \text{ iff } P(Y|B) > (a-b)/(u-u')(RR-1)$$

Thus (*) holds that one should consume A rather than B iff $P(Y|B) > (a - b)/(u - u')(RR - 1)$. This is enough to show that RR is not EU-sufficient, because the comparison of EU[A] with EU[B] requires not just RR and given costs and utilities, but also P(Y|B).

I will put this in slightly different terms. Note that any particular RR does not constrain the values that P(Y|B) can take, and neither do the values of a, b, u, or u', though obviously the value of P(Y|B) is mathematically bounded as all probabilities are (between 0 and 1). So, for any particular RR (wlog RR > 1) we can consider two cases:

$$\text{(i) } P(Y|B) = \left[(a-b)/(u-u')(RR-1) \right] - \varepsilon$$

$$\text{(ii) } P(Y|B) = \left[(a-b)/(u-u')(RR-1) \right] + \varepsilon$$

for some ε which is suitably small such that P(Y|B) remains bounded between 0 and 1, but which is suitably large such that:

in (i), $P(Y|B) < (a-b)/(u-u')(RR-1)$ and thus $EU[A] < EU[B]$,

and in (ii), $P(Y|B) > (a-b)/(u-u')(RR-1)$ and thus $EU[A] > EU[B]$.

Thus, when given a, b, u, and u', RR is insufficient to determine if the inequality in (*) holds. Thus RR is EU-insufficient.

A similar argument can be given to show that RRR is also EU-insufficient—see (Sprenger & Stegenga, forthcoming) for the proof.

Appendix 5: Modeling the Measurement of Effectiveness

The master argument for medical nihilism was based on the formal structure of Bayes' Theorem (§11.4). A central part of this argument involved employing this structure to model the influence of bias on one's confidence in the effectiveness of medical interventions. There is an alternative way to formally represent the influence of systematic bias on estimations of effectiveness. This follows an insight from Tal, who argues that measurement is 'model-based': measurement of a property occurs within a model of the measuring apparatus and the measured property (2016). I sketch such an approach here.

As I argue in Chapter 8 and Appendices 3 and 4, absolute measures of effectiveness, such as RD, are superior to relative measures, and so that is what I use here. I use RD_{trial} to represent the risk difference that is found in a trial. As I argue in §3.3, evidence from a trial bears on a hypothesis of the kind WORKS SOMEWHERE: the intervention was efficacious for a group of subjects in a trial. RD_{trial} is an *actual frequency*: that which was in fact measured in the trial. If $RD_{trial} > 0$, then we have evidence for a hypothesis of the kind WORKS SOMEWHERE. But as further argued in §3.3, hypotheses of the kind WORKS SOMEWHERE are not hypotheses about effectiveness; rather, they are hypotheses about efficacy, whereas effectiveness hypotheses are predictions about how an intervention will perform outside the controlled experimental setting, in the wild.

Effectiveness hypotheses can be of the kind WORKS GENERALLY: the intervention is effective for a population more general in a setting more varied than the controlled setting of a trial. I use RD_{target} to represent the risk difference that is expected when the intervention is used in the wild. RD_{target} is an *expected frequency*: that which we would expect to observe if the intervention were used in the wild. If $RD_{target} > 0$, then we have some evidence for a hypothesis of the kind WORKS GENERALLY.

As I further argue in §3.3 and §8.3, for a particular patient what matters are hypotheses of the kind WORKS FOR ME: the intervention will be effective for a particular patient. I use $RD_{patient}$ to represent the extent to which the intervention changes the probability that a patient will experience the beneficial outcome of interest. Abstracting much detail, this is a prediction about the future outcome of a single case, and thus $RD_{patient}$ is a *subjective probability*. It is this probability that is represented in many of the arguments in this book: the posterior probability in the master argument (§11.4), and the arguments in favor of RD (§8.3), including the epistemic argument (Appendix 3) and the decision-theoretic argument (Appendix 4).[1]

---

[1] Of course, the subjective probability in $RD_{patient}$ is not 'subjective' in the derogatory sense of being unconstrained by facts. It is in part determined by the objective expected frequency of RD_{target} (and so is also informed by the objective actual frequency of RD_{trial}), much like one's subjective belief that the next roll of a dice will be a six can be based at least in part on past experience of objective frequencies of sixes when rolling dice.

All measurement in science requires a chain of inferences, determined by one's model of the measuring instrument and its relation to the property being measured. To measure the effectiveness of a medical intervention, the chain of inferences is:

$$RD_{trial} \rightarrow RD_{target} \rightarrow RD_{patient}$$

The first inference, from RD_{trial} to RD_{target}, is an extrapolation (or generalization)—an inference from a measurement of a property performed in a controlled setting to an expectation about the value of that property in a more general target population. The second inference, from RD_{target} to $RD_{patient}$, is a particularization—an inference about the extent to which a particular person is like those people for whom the generalized inference of effectiveness was made (Fuller & Flores, 2015). I will focus on the first inference.

As noted in §8.4, many believe that the inference from RD_{trial} to RD_{target} is simple. That is, many assume that

$$RD_{target} = RD_{trial} \tag{1}$$

But, as argued in §8.4, (1) is generally false. We can add the concerns regarding bias in medical research articulated in Chapters 6, 8, 9, and 10 to the concerns about extrapolation articulated in §8.4 in order to develop a more sophisticated model of measuring RD_{target}. There are principled and empirical reasons that decisively indicate that the biases in medical research systematically overestimate the effectiveness of medical interventions. Taking these biases into account ought to lower our estimates of RD_{target} based on RD_{trial}. We can correct (1) in the same way that physicists correct their models of the dynamics of projectile objects when they take into account impeding forces such as friction. I will introduce terms to modify (1) that represent the influence that some of the biases that I have discussed throughout this book have on estimating RD_{target}: confirmation bias (which I denote C), instrument bias (I), recruitment bias (R), analysis bias (A), publication bias (P), and a single term to represent all other biases (O). Thus:

$$RD_{target} = RD_{trial} - C - I - R - A - P - O \tag{2}$$

Some of these biases are threats to internal validity while other are threats to external validity, but regardless, each of these biases systematically favor experimental drugs. The magnitudes of these correcting factors depend on details from case to case. But the directional influence is clear: each of these biases ought to lower our estimates of RD_{target} when extrapolating from RD_{trial}.

Given the widespread finding of small effect sizes in medical research (that is, RD_{trial} is tiny for many medical studies), the fact that RD_{target} must be even smaller than RD_{trial} (because of the correction factors associated with the various bias) is another way to articulate one important consideration in support of medical nihilism.

References

Alexandrova, A. (2008). First person reports and the measurement of happiness. *Philosophical Psychology, 21*(5), 571–83.

Alexandrova, A. (forthcoming). Can the science of well-being be objective? *The British Journal for the Philosophy of Science.*

Amundson, R. (2000). Against normal function. *Studies in the History and Philosophy of Biological and Biomedical Sciences, 31*(1), 33–53.

Angell, M. (2004a, July 15). The truth about the drug companies. *The New York Review of Books.*

Angell, M. (2004b). *The truth about the drug companies: how they deceive us and what to do about it.* New York: Random House.

Angell, M. (2009). Drug companies and doctors: a story of corruption. *The New York Review of Books.*

Ashcroft, R. (2002). What is clinical effectiveness? *Studies in the History and Philosophy of Biological and Biomedical Sciences, 33*(2), 219–33.

Assendelft, W. J., Koes, B. W., Knipschild, P. G., & Bouter, L. M. (1995). The relationship between methodological quality and conclusions in reviews of spinal manipulation. *JAMA, 274*(24), 1942–8.

Atkins, D., Best, D., Briss, P. A., & Group, G. W. (2004). Grading quality of evidence and strength of recommendations. *BMJ, 328*, 1490.

Bachand, A. M., Mundt, K. A., Mundt, D. J., & Montgomery, R. R. (2010). Epidemiological studies of formaldehyde exposure and risk of leukemia and nasopharyngeal cancer: a meta-analysis. *Crit Rev Toxicol, 40*(2), 85–100.

Bacon, F. (1605). The advancement of learning. In J. Spedding, R. Ellis, & D. Heath (eds.), *The works of Francis Bacon (1887–1901)* (Vol. 3).

Baigent, C., Keech, A., Kearney, P. M., Blackwell, L., Buck, G., Pollicino, C., . . . Simes, R. (2005). Efficacy and safety of cholesterol-lowering treatment: prospective meta-analysis of data from 90,056 participants in 14 randomised trials of statins. *Lancet, 366*(9493), 1267–78.

Balk, E. M., Bonis, P. A., Moskowitz, H., Schmid, C. H., Ioannidis, J. P., Wang, C., & Lau, J. (2002). Correlation of quality measures with estimates of treatment effect in meta-analyses of randomized controlled trials. *JAMA, 287*(22), 2973–82.

Barnes, D. E., & Bero, L. A. (1998). Why review articles on the health effects of passive smoking reach different conclusions. *JAMA, 279*(19), 1566–70.

Bartlett, C., Doyal, L., Ebrahim, S., Davey, P., Bachmann, M., Egger, M., & Dieppe, P. (2005). The causes and effects of socio-demographic exclusions from clinical trials. *Health Technol Assess, 9*(38), iii–iv, ix–x, 1–152.

Bass, A. (2008). *Side effects: a prosecutor, a whistleblower, and a bestselling antidepressant on trial.* Chapel Hill, NC: Algonquin Books.

Beasley, C. M., Jr., Dornseif, B. E., Bosomworth, J. C., Sayler, M. E., Rampey, A. H., Jr., Heiligenstein, J. H., . . . Masica, D. N. (1991). Fluoxetine and suicide: a meta-analysis of controlled trials of treatment for depression. *BMJ, 303*(6804), 685–92.

Bechtel, W., & Abrahamsen, A. (2005). Explanation: a mechanistic alternative. *Studies in the History and Philosophy of Biological and Biomedical Sciences, 36*, 421–41.

Becker, G. S. (2002, September 15). Get the FDA out of the way, and drug prices will drop. *Bloomberg Business.*

Berenson, A. (2006, December 17). Eli Lilly said to play down risk of top pill. *The New York Times.*

Bero, L., Oostvogel, F., Bacchetti, P., & Lee, K. (2007). Factors associated with findings of published trials of drug–drug comparisons: why some statins appear more efficacious than others. *PLoS Med, 4*(6), e184.

Biddle, J. (2007). Lessons from the Vioxx debacle: what the privatization of science can teach us about social epistemology. *Social Epistemology, 21*(1), 21–39.

Biddle, J. (2013). Institutionalizing dissent: a proposal for an adversarial system of pharmaceutical research. *Kennedy Inst Ethics J, 23*(4), 325–53.

Biddle, J. B. (2014). Can patents prohibit research? On the social epistemology of patenting and licensing in science. *Studies in History and Philosophy of Science Part A, 45*, 14–23.

Bigelow, J. (1835). *A discourse on self-limited disease.* Boston: Nathan Hale.

Black, D. M., Cummings, S. R., Karpf, D. B., Cauley, J. A., Thompson, D. E., Nevitt, M. C.,... Ensrud, K. E. (1996). Randomised trial of effect of alendronate on risk of fracture in women with existing vertebral fractures. Fracture Intervention Trial Research Group. *Lancet, 348*(9041), 1535–41.

Bluhm, R. (2005). From hierarchy to network: a richer view of evidence for evidence-based medicine. *Perspect Biol Med, 48*(4), 535–47.

Bluhm, R. (2011). Jeremy Howick: the philosophy of evidence-based medicine. *Theoretical Medicine and Bioethics, 32*(6), 423–7.

Blumenthal, D. K., & Garrison, J. C. (2011). Pharmaco-dynamics: molecular mechanisms of drug action. In L. L. Brunton, B. A. Chabner, & B. C. Knollmann (eds.), *Goodman & Gilman's The Pharmacological Basis of Therapeutics* (12th ed.). New York: McGraw Hill.

Bobbio, M., Demichelis, B., & Giustetto, G. (1994). Completeness of reporting trial results: effect on physicians' willingness to prescribe. *Lancet, 343*(8907), 1209–11.

Bombardier, C., Laine, L., Burgos-Vargas, R., Davis, B., Day, R., Ferraz, M. B.,... Weaver, A. (2006). Response to expression of concern regarding VIGOR study. *New England Journal of Medicine, 354*(11), 1196–9.

Bombardier, C., Laine, L., Reicin, A., Shapiro, D., Burgos-Vargas, R., Davis, B.,... Schnitzer, T. J. (2000). Comparison of upper gastrointestinal toxicity of rofecoxib and naproxen in patients with rheumatoid arthritis. *New England Journal of Medicine, 343*(21), 1520–8.

Boorse, C. (1977). Health as a theoretical concept. *Philosophy of Science, 44*(4), 542–73.

Borenstein, M., Hedges, L. V., Higgins, J. P. T., & Rothstein, H. R. (2009). *Introduction to meta-analysis.* Chichester: John Wiley and Sons.

Borgerson, K. (2008). Valuing and evaluating evidence in medicine. Ph.D., University of Toronto.

Borgerson, K. (2009). Valuing evidence: bias and the evidence hierarchy of evidence-based medicine. *Perspect Biol Med, 52*(2), 218–33.

Bosetti, C., McLaughlin, J. K., Tarone, R. E., Pira, E., & La Vecchia, C. (2008). Formaldehyde and cancer risk: a quantitative review of cohort studies through 2006. *Annals of Oncology, 19*(1), 29–43.

Bovens, L., & Hartmann, S. (2004). *Bayesian epistemology.* Oxford: Oxford University Press.

Broadbent, A. (2013). *Philosophy of epidemiology*. London: Palgrave Macmillan.

Brown, J. (2008). The community of science®. In M. Carrier, D. Howard, & J. Kourany (eds.), *The challenge of the social and the pressure of practice: science and values revisited*. Pittsburgh: University of Pittsburgh Press.

Brown, M. (2013). Values in science beyond underdetermination and inductive risk. *Philosophy of Science, 80*, 829–39.

Brown, S. (2016). *Through the valley of shadows: living wills, intensive care, and making medicine human*. New York: Oxford University Press.

Brownlee, S. (2008). *Overtreated: why too much medicine is making us sicker and poorer*. London: Bloomsbury.

Carey, B., & Harris, G. (2008, July 12). Psychiatric group faces scrutiny over drug industry ties. *The New York Times*.

Cartwright, N. (1979). Causal laws and effective strategies. *Nous, 13*, 419–37.

Cartwright, N. (2007). Are RCTs the gold standard? *BioSocieties, 2*(1), 11–20.

Cartwright, N. (2009). Evidence-based policy: what's to be done about relevance? *Philosophical Studies, 143*(1), 127–36.

Cartwright, N. (2010). What are randomised controlled trials good for? *Philosophical Studies, 147*, 59–70.

Cartwright, N. (2011). Evidence, external validity, and explanatory relevance. In Gregory J. Morgan (ed.), *Philosophy of science matters: the philosophy of Peter Achinstein*. New York: Oxford University Press.

Cartwright, N. (2012). Will this policy work for you? Predicting effectiveness better: how philosophy helps. *Philosophy of Science, 79*(5), 973–89.

Cassels, A. (2012). *Seeking sickness: medical screening and the misguided hunt for disease*. Vancouver: Greystone Books.

Chalmers, T. C., Smith, H., Jr., Blackburn, B., Silverman, B., Schroeder, B., Reitman, D., & Ambroz, A. (1981). A method for assessing the quality of a randomized control trial. *Control Clin Trials, 2*(1), 31–49.

Chang, H. (2004). *Inventing temperature*. New York: Oxford University Press.

Cho, M. K., & Bero, L. A. (1994). Instruments for assessing the quality of drug studies published in the medical literature. *JAMA, 272*(2), 101–4.

Clark, H. D., Wells, G. A., Huët, C., McAlister, F. A., Salmi, L. R., Fergusson, D., & Laupacis, A. (1999). Assessing the quality of randomized trials: reliability of the Jadad Scale. *Control Clin Trials, 20*(5), 448–52.

Cohen, J. (1960). A coefficient of agreement for nominal scales. *Educational and Psychological Measurement, 20*(1), 37–46.

Collins, J. J., & Lineker, G. A. (2004). A review and meta-analysis of formaldehyde exposure and leukemia. *Regul Toxicol Pharmacol, 40*(2), 81–91.

Cook, T. D., & Campbell, D. T. (1979). *Quasi-experimentation: design and analysis issues for field settings*. Boston: Houghton Mifflin.

Coombes, B. K., Bisset, L., & Vicenzino, B. (2010). Efficacy and safety of corticosteroid injections and other injections for management of tendinopathy: a systematic review of randomised controlled trials. *Lancet, 376*(9754), 1751–67.

Cooper, R. (2002). Disease. *Studies in the History and Philosophy of Biological and Biomedical Sciences, 33*, 263–82.

Cooper, R. (2013). Disease mongering. In H. La Follette (ed.), *The international encyclopaedia of ethics*. Oxford: Wiley-Blackwell.

Craver, C. (2007). *Explaining the brain: mechanisms and the mosaic unity of neuroscience*. New York: Oxford University Press.

Curfman, G. D., Morrissey, S., & Drazen, J. M. (2005). Expression of concern: Bombardier et al., "Comparison of upper gastrointestinal toxicity of rofecoxib and naproxen in patients with rheumatoid arthritis," N Engl J Med 2000;343:1520–8. *New England Journal of Medicine, 353*(26), 2813–14.

Daston, L., & Galison, P. (2007). *Objectivity*. New York: Zone Books.

Decullier, E., Chan, A.-W., & Chapuis, F. (2009). Inadequate dissemination of Phase I trials: a retrospective cohort study. *PLoS Med, 6*(2), e1000034.

de Melo-Martin, I., & Intemann, K. (2009). How do disclosure policies fail? Let us count the ways. *FASEB J, 23*(6), 1638–42.

de Melo-Martin, I., & Intemann, K. (2011). Feminist resources for biomedical research: lessons from the HPV vaccines. *Hypatia, 26*(1), 79–101.

Department of Clinical Epidemiology and Biostatistics, M. U. H. S. C. (1981). How to read clinical journals: V: To distinguish useful from useless or even harmful therapy. *Can Med Assoc J, 124*(9), 1156–62.

Descartes, R. (1637). *Discourse on the method*.

Doll, R. (2003). Fisher and Bradford Hill: their personal impact. *Int J Epidemiol, 32*(6), 929–31.

Doll, R., & Hill, A. B. (1950). Smoking and carcinoma of the lung; preliminary report. *Br Med J, 2*(4682), 739–48.

Doll, R., & Hill, A. B. (1954). The mortality of doctors in relation to their smoking habits: a preliminary report. *Br Med J, 1*(4877), 1451–5.

Donohue, J. M., Cevasco, M., & Rosenthal, M. B. (2007). A decade of direct-to-consumer advertising of prescription drugs. *New England Journal of Medicine, 357*(7), 673–81.

Doshi, P. (2009). Neuraminidase inhibitors: the story behind the Cochrane review. *BMJ, 339*, b5164.

Doshi, P., Jones, M., & Jefferson, T. (2012). Rethinking credible evidence synthesis. *BMJ, 344*, d7898.

Douglas, H. (2000). Inductive risk and values in science. *Philosophy of Science, 67*(4), 559–79.

Douglas, H. (2004). The irreducible complexity of objectivity. *Synthese, 138*(3), 453–73.

Douglas, H. (2009). *Science, policy, and the value-free ideal*. Pittsburgh: University of Pittsburgh Press.

Douglas, H. (2012). Weighing complex evidence in a democratic society. *Kennedy Inst Ethics J, 22*(2), 139–62.

Drews, J. (2004). Paul Ehrlich: magister mundi. *Nat Rev Drug Discov, 3*(9), 797–801.

Dutta, A. (2009). Discovery of new medicines. In J. P. Griffin (ed.), *The textbook of pharmaceutical medicine* (6th ed.). Oxford: Wiley-Blackwell.

Dwan, K., Altman, D. G., Arnaiz, J. A., Bloom, J., Chan, A. W., Cronin, E., . . . Williamson, P. R. (2008). Systematic review of the empirical evidence of study publication bias and outcome reporting bias. *PLoS One, 3*(8), e3081.

Earman, J. (1992). *Bayes or bust: a critical examination of Bayesian confirmation theory*. Cambridge, MA: MIT Press.

Ebrahim, S., Bance, S., Athale, A., Malachowski, C., & Ioannidis, J. P. (2016). Meta-analyses with industry involvement are massively published and report no caveats for antidepressants. *J Clin Epidemiol, 70,* 155–63.

Egger, M., Smith, G. D., & Phillips, A. N. (1997). Meta-analysis: principles and procedures. *BMJ, 315*(7121), 1533–7.

Elliott, C. (2010). *White coat, black hat: adventures on the dark side of medicine.* Boston: Beacon Press.

Elliott, K. C. (2011). Direct and indirect roles for values in science. *Philosophy of Science, 78*(2), 303–24.

Emanuel, E. (2015, November 21). Are good doctors bad for your health? *The New York Times.*

Epstein, S. (2007). *Inclusion: the politics of difference in medical research.* Chicago: Chicago University Press.

Ereshefsky, M. (2009). Defining 'health' and 'disease'. *Studies in the History and Philosophy of Biological and Biomedical Sciences, 40,* 221–7.

Eyding, D., Lelgemann, M., Grouven, U., Härter, M., Kromp, M., Kaiser, T., . . . Wieseler, B. (2010). Reboxetine for acute treatment of major depression: systematic review and meta-analysis of published and unpublished placebo and selective serotonin reuptake inhibitor controlled trials. *BMJ, 341:* c4737.

Fergusson, D., Doucette, S., Glass, K. C., Shapiro, S., Healy, D., Hebert, P., & Hutton, B. (2005). Association between suicide attempts and selective serotonin reuptake inhibitors: systematic review of randomised controlled trials. *BMJ, 330*(7488), 396.

Fleck, L. (1979). *Genesis and development of a scientific fact.* Chicago: University of Chicago Press.

Forrow, L., Taylor, W. C., & Arnold, R. M. (1992). Absolutely relative: how research results are summarized can affect treatment decisions. *Am J Med, 92*(2), 121–4.

Foucault, M. (1973). *The birth of the clinic: an archeology of medical perception.* London: Tavistock Publications Ltd.

Foucault, M. (2004 [1974]). The crisis of medicine or the crisis of antimedicine? *Foucault Studies, 1,* 5–19.

Fournier, J. C., DeRubeis, R. J., Hollon, S. D., Dimidjian, S., Amsterdam, J. D., Shelton, R. C., & Fawcett, J. (2010). Antidepressant drug effects and depression severity: a patient-level meta-analysis. *JAMA, 303*(1), 47–53.

Frances, A. (2015, January 12). The crisis of confidence in medical research. *Huffington Post.*

Frick, M. H., Elo, O., Haapa, K., Heinonen, O. P., Heinsalmi, P., Helo, P., et al. (1987). Helsinki Heart Study: primary-prevention trial with gemfibrozil in middle-aged men with dyslipidemia. Safety of treatment, changes in risk factors, and incidence of coronary heart disease. *New England Journal of Medicine, 317*(20), 1237–45.

Friedman, M., & Friedman, R. (1990). *Free to choose: a personal statement.* Boston: Mariner Books.

Friedman, R. (2009, May 18). New drugs have allure, not track record. *The New York Times.*

Frigerio, F., Casimir, M., Carobbio, S., & Maechler, P. (2008). Tissue specificity of mitochondrial glutamate pathways and the control of metabolic homeostasis. *Biochimica et Biophysica Acta (BBA)—Bioenergetics, 1777*(7–8), 965–72.

Frobell, R. B., Roos, E. M., Roos, H. P., Ranstam, J., & Lohmander, L. S. (2010). A randomized trial of treatment for acute anterior cruciate ligament tears. *New England Journal of Medicine, 363*(4), 331–42.

Fuller, J. (2013a). Rationality and the generalization of randomized controlled trial evidence. *J Eval Clin Pract, 19*, 644–7.

Fuller, J. (2013b). Rhetoric and argumentation: how clinical practice guidelines think. *J Eval Clin Pract, 19*, 433–41.

Fuller, J. (forthcoming). The confounding question of confounding causes in randomized trials. *The British Journal for the Philosophy of Science.*

Fuller, J., & Flores, L. J. (2015). The Risk GP model: the standard model of prediction in medicine. *Studies in History and Philosophy of Biological and Biomedical Sciences, 54*, 49–61.

Gabriel, J. (2014). *Medical monopoly: intellectual property rights and the origins of the modern pharmaceutical industry.* Chicago: University of Chicago Press.

Gambacorti-Passerini, C., Antolini, L., Mahon, F.-X., Guilhot, F., Deininger, M., Fava, C.,… Kim, D.-W. (2011). Multicenter independent assessment of outcomes in chronic myeloid leukemia patients treated with imatinib. *J Natl Cancer Inst, 103*(7), 553–61.

Garfinkel, D., & Mangin, D. (2010). Feasibility study of a systematic approach for discontinuation of multiple medications in older adults: addressing polypharmacy. *Arch Intern Med, 170*(18), 1648–54.

Ghaemi, N. (2012). Taking disease seriously: beyond 'pragmatic' nosology. In K. S. Kendler & J. Parnas (eds.), *Philosophical issues in psychiatry*, Volume 2: *Nosology.* Oxford: Oxford University Press.

Glass, G. (1976). Primary, secondary, and meta-analysis of research. *Educational Researcher, 5*(10), 3–8.

Godfrey-Smith, P. (2009). *Darwinian populations and natural selection.* Oxford: Oxford University Press.

Goldacre, B. (2012). *Bad pharma: how drug companies mislead doctors and harm patients.* London: HarperCollins.

Goldenberg, M. J. (2009). Iconoclast or creed? Objectivism, pragmatism, and the hierarchy of evidence. *Perspect Biol Med, 52*(2), 168–87.

González-Moreno, M., Saborido, C., & Teira, D. (2015). Disease-mongering through clinical trials. *Studies in History and Philosophy of Biological and Biomedical Sciences, 51*, 11–18.

Gøtzsche, P. (2013). *Deadly medicines and organized crime.* London: Radcliffe Publishing.

Graham, D. (2005). FDA incapable of protecting U.S., scientist alleges/Interviewer: D. Carozza. *Fraud Magazine.*

Gunnell, D., Saperia, J., & Ashby, D. (2005). Selective serotonin reuptake inhibitors (SSRIs) and suicide in adults: meta-analysis of drug company data from placebo controlled, randomised controlled trials submitted to the MHRA's safety review. *BMJ, 330*(7488), 385.

Gurwitz, J. H., Col, N. F., & Avorn, J. (1992). The exclusion of the elderly and women from clinical trials in acute myocardial infarction. *JAMA, 268*(11), 1417–22.

Guyatt, G., & Rennie, D. (2001). *User's guide to the medical literature.* Chicago: AMA Press.

Hacking, I. (1988). Telepathy: origins of randomization in experimental design. *Isis, 79*, 427–51.

Hadorn, D. C., Baker, D., Hodges, J. S., & Hicks, N. (1996). Rating the quality of evidence for clinical practice guidelines. *J Clin Epidemiol, 49*, 749–54.

Hamilton, M. (1960). A rating scale for depression. *J Neurol Neurosurg Psychiat, 23*, 56–62.

Harris, G. (2008, October 3). Top psychiatrist didn't report drug makers' pay. *The New York Times.*

Harris, G. (2010, February 22). A face-off on the safety of a drug for diabetes. *The New York Times*.

Hartling, L., Bond, K., Vandermeer, B., Seida, J., Dryden, D. M., & Rowe, B. H. (2011). Applying the risk of bias tool in a systematic review of combination long-acting beta-agonists and inhaled corticosteroids for persistent asthma. *PLoS One, 6*(2), e17242.

Hartling, L., Ospina, M., Liang, Y., Dryden, D. M., Hooton, N., Krebs Seida, J., & Klassen, T. P. (2009). Risk of bias versus quality assessment of randomised controlled trials: cross sectional study. *BMJ, 339*, b4012.

Hausman, D. (2012). Health, naturalism, and functional efficiency. *Philosophy of Science, 79*, 519–41.

Hazell, L., & Shakir, S. A. (2006). Under-reporting of adverse drug reactions: a systematic review. *Drug Saf, 29*(5), 385–96.

Hempel, S., Suttorp, M. J., Miles, J. N. V., Wang, Z., Maglione, M., Morton, S.,...Shekelle, P. G. (2011). Empirical evidence of associations between trial quality and effect size. In *AHRQ methods for effective health care*. Rockville, MD: Agency for Healthcare Research and Quality (US).

Herbison, P., Hay-Smith, J., & Gillespie, W. J. (2006). Adjustment of meta-analyses on the basis of quality scores should be abandoned. *J Clin Epidemiol, 59*(12), 1249–56.

Hesslow, G. (1993). Do we need a concept of disease? *Theoretical Medicine and Bioethics, 14*, 1–14.

Holman, B. (2015). Why most sugar pills are not placebos. *Philosophy of Science, 82*(5), 1330–43.

Holmes, O. W., Sr. (1860). Annual meeting of the Massachusetts Medical Society (May 30, 1860).

Horton, R. (2000). Genetically modified food: consternation, confusion, and crack-up. *Med J Aust, 172*(4), 148–9.

Horton, R. (2001). Lotronex and the FDA: a fatal erosion of integrity. *The Lancet, 357*(9268), 1544–5.

Horton, R. (2015). Offline: what is medicine's 5 sigma? *The Lancet, 385*, 1380.

Horwitz, A., & Wakefield, J. (2007). *Loss of sadness: how psychiatry transformed normal sorrow into depressive disorder*. New York: Oxford University Press.

Howick, J. (2011a). Exposing the vanities—and a qualified defense—of mechanistic reasoning in health care decision making. *Philosophy of Science, 78*(5), 926–40.

Howick, J. (2011b). *The philosophy of evidence-based medicine*. Oxford: Wiley-Blackwell.

Howick, J., Glasziou, P., & Aronson, J. K. (2009). The evolution of evidence hierarchies: what can Bradford Hill's 'guidelines for causation' contribute? *J R Soc Med, 102*(5), 186–94.

Howson, C., & Urbach, P. (1989). *Scientific reasoning: the Bayesian approach*. La Salle, IL: Open Court.

Hoyningen-Huene, P. (2013). *Systematicity: the nature of science*. Oxford: Oxford University Press.

Illari, P. M. (2011). Mechanistic evidence: disambiguating the Russo–Williamson thesis. *International Studies in the Philosophy of Science, 25*(2), 139–57.

Illich, I. (1975). *Medical nemesis: the expropriation of health*. London: Marion Boyars.

Ioannidis, J. P. (2005a). Contradicted and initially stronger effects in highly cited clinical research. *JAMA, 294*(2), 218–28.

Ioannidis, J. P. (2005b). Why most published research findings are false. *PLoS Med, 2*(8), e124.

Ioannidis, J. P. (2008a). Effectiveness of antidepressants: an evidence myth constructed from a thousand randomized trials? *Philos Ethics Humanit Med, 3*, 14.

Ioannidis, J. P. (2008b). Why most discovered true associations are inflated. *Epidemiology, 19*(5), 640–8.

Ioannidis, J. P. (2011). An epidemic of false claims: competition and conflicts of interest distort too many medical findings. *Sci Am, 304*(6), 16.

Jaber, B. L., Lau, J., Schmid, C. H., Karsou, S. A., Levey, A. S., & Pereira, B. J. (2002). Effect of biocompatibility of hemodialysis membranes on mortality in acute renal failure: a meta-analysis. *Clin Nephrol, 57*(4), 274–82.

Jadad, A. R., Moore, R. A., Carroll, D., Jenkinson, C., Reynolds, D. J., Gavaghan, D. J., & McQuay, H. J. (1996). Assessing the quality of reports of randomized clinical trials: is blinding necessary? *Control Clin Trials, 17*(1), 1–12.

James, J. T. (2013). A new, evidence-based estimate of patient harms associated with hospital care. *Journal of Patient Safety, 9*(3), 122–8.

Jefferson, T., Jones, M., Doshi, P., & Del Mar, C. (2009). Neuraminidase inhibitors for preventing and treating influenza in healthy adults: systematic review and meta-analysis. *BMJ, 339*, b5106.

Jefferson, T., Jones, M. A., Doshi, P., Del Mar, C. B., Heneghan, C. J., Hama, R., & Thompson, M. J. (2012). Neuraminidase inhibitors for preventing and treating influenza in healthy adults and children. *Cochrane Database Syst Rev, 1*, CD008965.

Jena, A. B., Prasad, V., Goldman, D. P., & Romley, J. (2015). Mortality and treatment patterns among patients hospitalized with acute cardiovascular conditions during dates of national cardiology meetings. *JAMA Internal Medicine, 175*(2), 237–44.

Jennings, C. (2006). What you can't measure, you can't manage: the need for quantitative indicators in peer review. *Nature*. doi:10.1038/nature05032.

Jensen, P. S., Arnold, L. E., Swanson, J. M., Vitiello, B., Abikoff, H. B., Greenhill, L. L., ... Hur, K. (2007). 3-year follow-up of the NIMH MTA study. *J Am Acad Child Adolesc Psychiatry, 46*(8), 989–1002.

Jukola, S. (2015). Meta-analysis, ideals of objectivity, and the reliability of medical knowledge. *Science and Technology Studies, 28*(3), 101–21.

Jüni, P., Witschi, A., Bloch, R., & Egger, M. (1999). The hazards of scoring the quality of clinical trials for meta-analysis. *JAMA, 282*(11), 1054–60.

Kaiser, L., Wat, C., Mills, T., Mahoney, P., Ward, P., & Hayden, F. (2003). Impact of oseltamivir treatment on influenza-related lower respiratory tract complications and hospitalizations. *Arch Intern Med, 163*(14), 1667–72.

Kaplan, R. M., & Irvin, V. L. (2015). Likelihood of null effects of large NHLBI clinical trials has increased over time. *PLoS One, 10*(8), e0132382.

Karanicolas, P. J., Kunz, R., & Guyatt, G. H. (2008). Point: evidence-based medicine has a sound scientific base. *Chest, 133*(5), 1067–71.

Kelly, M. P., & Moore, T. A. (2011). The judgement process in evidence-based medicine and health technology assessment. *Social Theory & Health, 10*(1), 1–19.

Kendler, K. S. (2012). Introduction to Chapter 8. In K. Kendler & J. Parnas (eds.), *Philosophical issues in psychiatry*, Volume 2: *Nosology*. Oxford: Oxford University Press.

King, N. B., Harper, S., & Young, M. E. (2012). Use of relative and absolute effect measures in reporting health inequalities: structured review. *BMJ, 345*: e5774.

Kingma, E. (2007). What is it to be healthy? *Analysis, 67*(294), 128–33.

Kirsch, I. (2011). *The emperor's new drugs: exploding the antidepressant myth*. New York: Basic Books.

Kirsch, I., Deacon, B. J., Huedo-Medina, T. B., Scoboria, A., Moore, T. J., & Johnson, B. T. (2008). Initial severity and antidepressant benefits: a meta-analysis of data submitted to the Food and Drug Administration. *PLoS Med, 5*(2), e45.

Kirsch, I., Moore, T. J., Scoboria, A., & Nicholls, S. S. (2002). The emperor's new drugs: an analysis of antidepressant medication data submitted to the US Food and Drug Administration. *Prevention & Treatment, 5*(1), 23a.

Kitano, H. (2007). A robustness-based approach to systems-oriented drug design. *Nat Rev Drug Discov, 6*(3), 202–10.

Kitcher, P. (2011). *Science in a democratic society*. New York: Prometheus Books.

Knipschild, P. (1994). Systematic reviews: some examples. *BMJ, 309*(6956), 719–21.

Koehler, J. J. (1993). The influence of prior beliefs on scientific judgments of evidence quality. *Organizational Behavior and Human Decision Processes, 56*(1), 28–55.

Kolata, G. (2009, April 23). Advances elusive in the drive to cure cancer. *The New York Times*.

Krimsky, S. (2003). *Science in the private interest*. Lanham, MD: Rowman & Littlefield.

Krueger, J. (2015). Theoretical health and medical practice. *Philosophy of Science, 82*(3), 491–508.

Kuhn, T. (1962). *The Structure of scientific revolutions*. Chicago: University of Chicago Press.

Kukla, R. (2012). "Author TBD": radical collaboration in contemporary biomedical research. *Philosophy of Science, 79*, 845–58.

Kunda, Z. (1990). The case for motivated reasoning. *Psychol Bull, 108*(3), 480–98.

La Caze, A. (2011). The role of basic science in evidence-based medicine. *Biology and Philosophy, 26*(1), 81–98.

Laing, R. D. (2011 [1968]). *The politics of the family*. Toronto: House of Anansi Press.

Lange, M. (2007). The end of diseases. *Philosophical Topics, 35*(1&2), 265–92.

Laughren, T. P. (2004). Background comments for February 2, 2004. Meeting of Psychopharmacological Drugs Advisory Committee (PDAC) and Pediatric Subcommittee of the AntiInfective Drugs Advisory Committee (Peds AC). Retrieved from http://www.fda. gov/ohrms/dockets/ac/04/briefing/4006B1_03_backgroundmemo01-05-04.pdf.

Lavery, A. M., Verhey, L. H., & Waldman, A. T. (2014). Outcome measures in relapsing-remitting multiple sclerosis: capturing disability and disease progression in clinical trials. *Multiple Sclerosis International*, Article ID 262350, http://dx.doi.org/10.1155/2014/262350.

Lemmens, T., & Telfer, C. (2012). Access to information and the right to health: the human rights case for clinical trials transparency. *American Journal of Law and Medicine, 38*, 63–112.

Lemoine, M. (2013). Defining disease beyond conceptual analysis: an analysis of conceptual analysis in philosophy of medicine. *Theor Med Bioeth, 34*(4), 309–25.

Lemoine, M. (2017). Animal extrapolation in preclinical studies: an analysis of the tragic case of TGN1412. *Studies in History and Philosophy of Biological and Biomedical Sciences, 61*, 35–45.

Leuridan, B., & Weber, E. (2011). The IARC and mechanistic evidence. In P. M. Illari, F. Russo, & J. Williamson (eds.), *Causality in the sciences*. Oxford: Oxford University Press.

Linde, K., Clausius, N., Ramirez, G., Melchart, D., Eitel, F., Hedges, L. V., & Jonas, W. B. (1997). Are the clinical effects of homeopathy placebo effects? A meta-analysis of placebo-controlled trials. *Lancet, 350*(9081), 834–43.

Longino, H. (1990). *Science as social knowledge: values and objectivity in scientific inquiry.* Princeton: Princeton University Press.

Lurie, P., & Wolfe, S. (1998). *FDA medical officers report lower standards permit dangerous drug approvals.* Retrieved from https://www.citizen.org/our-work/health-and-safety/fda-medical-officers-report-lower-standards-permit-dangerous.

Machamer, P., Darden, L., & Craver, C. (2000). Thinking about mechanisms. *Philosophy of Science, 67*(1), 1–25.

Malenka, D., Baron, J., Johansen, S., Wahrenberger, J., & Ross, J. (1993). The framing effect of relative and absolute risk. *J Gen Intern Med, 8*(10), 543–8.

Malhotra, A., Maughan, D., Ansell, J., Lehman, R., Henderson, A., Gray, M.,...Bailey, S. (2015). Choosing wisely in the UK: the Academy of Medical Royal Colleges' initiative to reduce the harms of too much medicine. *BMJ, 350:* h2308.

Marmot, M. (2004). *Status syndrome: how your social standing directly affects your health and life expectancy.* London: Bloomsbury.

Mayo, D. (1996). *Error and the growth of experimental knowledge.* Chicago: University of Chicago Press.

McClimans, L. (2010). A theoretical framework for patient-reported outcome measures. *Theoretical Medicine and Bioethics, 31,* 225–40.

McGoey, L., & Jackson, E. (2009). Seroxat and the suppression of clinical trial data: regulatory failure and the uses of legal ambiguity. *Journal of Medical Ethics, 35*(2), 107–12.

McKeown, T. (1976a). *The modern rise of population.* London: Edward Arnold.

McKeown, T. (1976b). *The role of medicine: dream, mirage, or nemesis?* London: Nuffield Provincial Hospitals Trust.

McKeown, T., & Record, R. G. (1962). Reasons for the decline of mortality in England and Wales during the nineteenth century. *Population Studies, 16*(2), 94–122.

Medawar, P. (1983, 1 December). The pissing evile. *London Review of Books, 5,* 10–11.

Miller, B. (2010). *A social theory of knowledge.* Ph.D., University of Toronto.

Millikan, R. G. (1989). In defense of proper functions. *Philosophy of Science, 56*(2), 288–302.

Mitchell, S. (2009). *Unsimple truths: science, complexity, and policy.* Chicago: University of Chicago Press.

Moher, D., Hopewell, S., Schulz, K. F., Montori, V., Gøtzsche, P. C., Devereaux, P. J.,...Altman, D. G. (2010). CONSORT 2010 explanation and elaboration: updated guidelines for reporting parallel group randomised trials. BMJ, 340: c869.

Moher, D., Jadad, A. R., Nichol, G., Penman, M., Tugwell, P., & Walsh, S. (1995). Assessing the quality of randomized controlled trials: an annotated bibliography of scales and checklists. *Control Clin Trials, 16*(1), 62–73.

Moher, D., Jadad, A. R., & Tugwell, P. (1996). Assessing the quality of randomized controlled trials: current issues and future directions. *Int J Technol Assess Health Care, 12*(2), 195–208.

Moher, D., Pham, B., Jones, A., Cook, D. J., Jadad, A. R., Moher, M.,...Klassen, T. P. (1998). Does quality of reports of randomised trials affect estimates of intervention efficacy reported in meta-analyses? *Lancet, 352*(9128), 609–13.

Molina, B. S., Hinshaw, S. P., Swanson, J. M., Arnold, L. E., Vitiello, B., Jensen, P. S.,...Houck, P. R. (2009). The MTA at 8 years: prospective follow-up of children treated for combined-type ADHD in a multisite study. *J Am Acad Child Adolesc Psychiatry, 48*(5), 484–500.

Moncrieff, J. (2013). *The bitterest pills: the troubling story of antipsychotic drugs.* New York: Palgrave Macmillan.

Moynihan, R. (2002). Alosetron: a case study in regulatory capture, or a victory for patients' rights? *BMJ, 325*(7364), 592–5.

Moynihan, R., & Cassels, A. (2005). *Selling sickness: how the world's biggest pharmaceutical companies are turning us all into patients.* Vancouver: Greystone Books.

Moynihan, R., Cooke, G. P. E., Doust, J. A., Bero, L., Hill, S., & Glasziou, P. P. (2013). Expanding disease definitions in guidelines and expert panel ties to industry: a cross-sectional study of common conditions in the United States. *PLoS Med, 10*(8), e1001500.

Moynihan, R., & Smith, R. (2002). Too much medicine? Almost certainly. *BMJ, 324*(7342), 859–60.

MTA. (1999). A 14-month randomized clinical trial of treatment strategies for attention-deficit/hyperactivity disorder. The MTA Cooperative Group. Multimodal Treatment Study of Children with ADHD. *Arch Gen Psychiatry, 56*(12), 1073–86.

Murphy, D. (2006). *Psychiatry in the scientific image.* Cambridge, MA: MIT Press.

Murphy, D. (2008). Health and disease. In S. Sarkar & A. Plutynski (eds.), *A companion to the philosophy of biology.* Malden, MA: Blackwell.

Naci, H., & Ioannidis, J. P. A. (2013). Comparative effectiveness of exercise and drug interventions on mortality outcomes: metaepidemiological study. *BMJ, 347*: f5577.

Naylor, C. D., Chen, E., & Strauss, B. (1992). Measured enthusiasm: does the method of reporting trial results alter perceptions of therapeutic effectiveness? *Ann Intern Med, 117*(11), 916–21.

Neander, K. (1991). Functions as selected effects: the conceptual analyst's defense. *Philosophy of Science, 58*, 168–84.

Nemeroff, C. B., Heim, C. M., Thase, M. E., Klein, D. N., Rush, A. J., Schatzberg, A. F.,…Keller, M. B. (2003). Differential responses to psychotherapy versus pharmacotherapy in patients with chronic forms of major depression and childhood trauma. *Proc Natl Acad Sci USA, 100*(24), 14293–6.

Nexøe, J., Gyrd-Hansen, D., Kragstrup, J., Kristiansen, I. S., & Nielsen, J. B. (2002). Danish GPs' perception of disease risk and benefit of prevention. *Fam Pract, 19*(1), 3–6.

Nissen, S. E. (2010). The rise and fall of rosiglitazone. *Eur Heart J, 31*(7), 773–6.

Nissen, S. E., & Wolski, K. (2007). Effect of rosiglitazone on the risk of myocardial infarction and death from cardiovascular causes. *New England Journal of Medicine, 356*(24), 2457–71.

NSF. (2014). Science and engineering indicators. Retrieved from http://www.nsf.gov/statistics/seind14/.

O'Connor, C. M., Starling, R. C., Hernandez, A. F., Armstrong, P. W., Dickstein, K., Hasselblad, V.,…Califf, R. M. (2011). Effect of nesiritide in patients with acute decompensated heart failure. *New England Journal of Medicine, 365*(1), 32–43.

Olivo, S. A., Macedo, L. G., Gadotti, I. C., Fuentes, J., Stanton, T., & Magee, D. J. (2008). Scales to assess the quality of randomized controlled trials: a systematic review. *Phys Ther, 88*(2), 156–75.

Osimani, B. (2014). Hunting side effects and explaining them: should we reverse evidence hierarchies upside down? *Topoi, 33*(2), 295–312.

Papanikolaou, P. N., Christidi, G. D., & Ioannidis, J. P. A. (2006). Comparison of evidence on harms of medical interventions in randomized and nonrandomized studies. *Can Med Assoc J, 174*(5), 635–41.

Papanikolaou, P. N., Churchill, R., Wahlbeck, K., & Ioannidis, J. P. (2004). Safety reporting in randomized trials of mental health interventions. *Am J Psychiatry, 161*(9), 1692–7.

Papineau, D. (1994). The virtues of randomization. *The British Journal for the Philosophy of Science, 45*(2), 437–50.

Patz, E. F., Jr., Pinsky, P., Gatsonis, C., Sicks, J. D., Kramer, B. S., Tammemagi, M. C., ... Aberle, D. R. (2014). Overdiagnosis in low-dose computed tomography screening for lung cancer. *JAMA Intern Med, 174*(2), 269–74.

Pauling, L. (1986). *How to live longer and feel better*. New York: W. H. Freeman & Co.

Payer, L. (1992). *Disease-mongers: how doctors, drug companies, and insurers are making you feel sick*. New York: Wiley.

Pearl, J. (2009). *Causality: models, reasoning, and inference* (2nd ed.). Cambridge: Cambridge University Press.

Pitrou, I., Boutron, I., Ahmad, N., & Ravaud, P. (2009). Reporting of safety results in published reports of randomized controlled trials. *Arch Intern Med, 169*(19), 1756–61.

Plutynski, A. (2017). Safe, or sorry? Cancer screening and inductive risk. In K. Elliot & T. Richards (eds.), *Exploring Inductive Risk*. New York: Oxford University Press.

Pogge, T. W. (2005). Human rights and global health: a research program. *Metaphilosophy, 36*(1–2), 182–209.

Popper, K. (1959 [1935]). *The logic of scientific discovery*. London: Hutchinson.

Porter, R. (1999). *The greatest benefit to mankind: a medical history of humanity*. New York: W. W. Norton & Co.

Post, P. N., de Beer, H., & Guyatt, G. H. (2013). How to generalize efficacy results of randomized trials: recommendations based on a systematic review of possible approaches. *J Eval Clin Pract, 19*(4), 638–43.

Pray, L. (2008). Gleevec: the breakthrough in cancer treatment. *Nature Education, 1*(1), 37.

Rawlins, M. (2008). *De testimonio: on the evidence for decisions about the use of therapeutic interventions*. London: Royal College of Physicians.

Regier, D. A. (2012). Diagnostic threshold considerations for DSM-5. In K. S. Kendler & J. Parnas (eds.), *Philosophical issues in psychiatry*, Volume 2: *Nosology*. Oxford: Oxford University Press.

Reginster, B. (2006). *The affirmation of life: Nietzsche on overcoming nihilism*. Cambridge, MA: Harvard University Press.

Reisch, J. S., Tyson, J. E., & Mize, S. G. (1989). Aid to the evaluation of therapeutic studies. *Pediatrics, 84*(5), 815–27.

Reiss, J., & Kitcher, P. (2009). Biomedical research, neglected diseases, and well-ordered science. *Theoria, 24*(3), 263–82.

Resnik, D. (2007). *The price of truth: how money affects the norms of science*. New York: Oxford University Press.

Rhine, J. B., Pratt, J. G., Stuart, C. E., Smith, B. M., & Greenwood, J. A. (1940). *Extrasensory perception after sixty years*. New York: Holt.

Robertson, W. (1976, March 1976). Merck strains to keep the pots aboiling. *Fortune*.

Rothman, K. J., & Greenland, S. (2005). Causation and causal inference in epidemiology. *Am J Public Health, 95 Suppl 1*, S144–50.

Rothwell, P. M. (2005). External validity of randomised controlled trials: "to whom do the results of this trial apply?". *Lancet, 365*(9453), 82–93.

Royall, R. (1997). *Statistical evidence: a likelihood paradigm*. London: Chapman & Hall.

Rudner, R. (1953). The scientist *qua* scientist makes value judgements. *Philosophy of Science, 20*, 1–6.

Russo, F., & Williamson, J. (2007). Interpreting causality in the health sciences. *International Studies in the Philosophy of Science, 21*, 157–70.

Sackett, D. L. (1979). Bias in analytic research. *Journal of Chronic Diseases, 32*(1–2), 51–63.

Saquib, N., Saquib, J., & Ioannidis, J. P. (2015). Does screening for disease save lives in asymptomatic adults? Systematic review of meta-analyses and randomized trials. *International Journal of Epidemiology, 44*(1), 264–77.

Schachter, H. M., Pham, B., King, J., Langford, S., & Moher, D. (2001). How efficacious and safe is short-acting methylphenidate for the treatment of attention-deficit disorder in children and adolescents? A meta-analysis. *Can Med Assoc J, 165*(11), 1475–88.

Schaffner, K. (1993). *Discovery and explanation in biology and medicine.* Chicago: University of Chicago Press.

Scheen, A. J. (2012). Outcomes and lessons from the PROactive study. *Diabetes Res Clin Pract, 98*(2), 175–86.

Schwartz, A. (2015, December 10). Psychiatric drugs are being prescribed to infants. *The New York Times.*

Schwartz, L. M., & Woloshin, S. (2011). Communicating uncertainties about prescription drugs to the public: a national randomized trial. *Arch Intern Med, 171*(16), 1463–8.

Schwartz, P. (2007). Defining dysfunction: natural selection, design, and drawing a line. *Philosophy of Science, 74*, 364–85.

Schwartz, P. H., & Meslin, E. M. (2008). The ethics of information: absolute risk reduction and patient understanding of screening. *J Gen Intern Med, 23*(6), 867–70.

Slater, L. (1999). *Prozac diary.* London: Penguin Books.

Slavin, R. E. (1995). Best evidence synthesis: an intelligent alternative to meta-analysis. *J Clin Epidemiol, 48*(1), 9–18.

Sober, E. (2008). *Evidence and evolution: the logic behind the science.* Cambridge: Cambridge University Press.

Sober, E. (2009). Absence of evidence and evidence of absence: evidential transitivity in connection with fossils. *Philosophical Studies, 143*(1), 63–90.

Solomon, M. (2001). *Social empiricism.* Cambridge, MA: MIT Press.

Solomon, M. (2011). Just a paradigm: evidence-based medicine in epistemological context. *European Journal for Philosophy of Science, 1*(3), 451–66.

Sorensen, L., Gyrd-Hansen, D., Kristiansen, I. S., Nexoe, J., & Nielsen, J. B. (2008). Laypersons' understanding of relative risk reductions: randomised cross-sectional study. *BMC Med Inform Decis Mak, 8*, 31.

Spector, R. (2010). A skeptic's view of pharmaceutical progress. *Skeptical Inquirer, 34.*

Spielmans, G. I., & Kirsch, I. (2014). Drug approval and drug effectiveness. *Annual Review of Clinical Psychology, 10*(1), 741–66.

Spitzer, W. O., Lawrence, V., Dales, R., Hill, G., Archer, M. C., Clark, P., ... et al. (1990). Links between passive smoking and disease: a best-evidence synthesis. A report of the Working Group on Passive Smoking. *Clin Invest Med, 13*(1), 17–42; discussion 43–16.

Sprenger, J., & Stegenga, J. (forthcoming). Three arguments for absolute outcome measures. *Philosophy of Science.*

Steel, D. (2007). *Across the boundaries.* New York: Oxford University Press.

Steel, D. (2010). Epistemic values and the argument from inductive risk. *Philosophy of science, 77*(1), 14–34.

Steele, K. (2012). The scientist qua policy advisor makes value judgments. *Philosophy of science,* *79*(5), 893–904.

Stegenga, J. (2011). Is meta-analysis the platinum standard? *Studies in History and Philosophy of Biological and Biomedical Sciences, 42,* 497–507.

Stegenga, J. (2014). Down with the hierarchies. *Topoi, 33*(2), 313–22.

Stegenga, J. (2015a). Effectiveness of medical interventions. *Studies in History and Philosophy of Biological and Biomedical Sciences, 54,* 34–44.

Stegenga, J. (2015b). Herding QATs: quality assessment tools for evidence in medicine. In P. Huneman, G. Lambert, & M. Silberstein (eds.), *Classification, disease and evidence: new essays in the philosophy of medicine.* Dordrecht: Springer Netherlands.

Stegenga, J. (2015c). Measuring effectiveness. *Studies in History and Philosophy of Biological and Biomedical Sciences, 54,* 62–71.

Stegenga, J. (2016). Hollow hunt for harms. *Perspectives on Science, 24*(5), 481–504.

Stegenga, J. (2017). Drug regulation and the inductive risk calculus. In K. Elliot & T. Richards (eds.), *Exploring inductive risk.* New York: Oxford University Press.

Straus, S. E., Richardson, W. S., Glasziou, P. P., & Haynes, R. B. (2005). *Evidence-based medicine: how to practice and teach* (3rd ed.). London: Elsevier Churchill Livingstone.

Strevens, M. (2009). Objective evidence and absence: comment on Sober. *Philosophical Studies, 143,* 91–100.

Subramanian, S., Venkataraman, R., & Kellum, J. A. (2002). Influence of dialysis membranes on outcomes in acute renal failure: a meta-analysis. *Kidney Int, 62*(5), 1819–23.

Suppes, P., & Zinnes, J. L. (1962). Basic measurement theory. *Institute for Mathematical Studies in the Social Sciences, Technical Report No. 45.*

Swanson, J. M., Elliott, G. R., Greenhill, L. L., Wigal, T., Arnold, L. E., Vitiello, B., … Volkow, N. D. (2007). Effects of stimulant medication on growth rates across 3 years in the MTA follow-up. *J Am Acad Child Adolesc Psychiatry, 46*(8), 1015–27.

Szklo, M., & Nieto, J. (2007). *Epidemiology: beyond the basics* (2nd ed.). Boston: Jones and Bartlett Publishers.

Tabery, J. (2014). *Beyond versus: the struggle to understand the interaction of nature and nurture.* Cambridge, MA: MIT Press.

Tal, E. (2011). How accurate is the standard second? *Philosophy of Science, 78*(5), 1082–96.

Tal, E. (2016). Making time: a study in the epistemology of measurement. *The British Journal for the Philosophy of Science, 67,* 297–335.

Teller, P. (2013). The concept of measurement-precision. *Synthese, 190,* 189–202.

Thagard, P. (2003). Pathways to biomedical discovery. *Philosophy of Science, 70,* 235–54.

Thalos, M. (2013). *Without hierarchies: the scale freedom of the universe.* New York: Oxford University Press.

Tricco, A. C., Tetzlaff, J., Pham, B., Brehaut, J., & Moher, D. (2009). Non-Cochrane vs. Cochrane reviews were twice as likely to have positive conclusion statements: cross-sectional study. *J Clin Epidemiol, 62*(4), 380–6 e381.

Tsang, R., Colley, L., & Lynd, L. D. (2009). Inadequate statistical power to detect clinically significant differences in adverse event rates in randomized controlled trials. *J Clin Epidemiol, 62*(6), 609–16.

Tsou, J. Y. (2012). Intervention, causal reasoning, and the neurobiology of mental disorders: pharmacological drugs as experimental instruments. *Studies in History and Philosophy of Biological and Biomedical Sciences, 43*(2), 542–51.

Tyrer, P., & Kendall, T. (2009). The spurious advance of antipsychotic drug therapy. *Lancet, 373*(9657), 4–5.

Upshur, R. E. (2005). Looking for rules in a world of exceptions: reflections on evidence-based practice. *Perspect Biol Med, 48*(4), 477–89.

van Fraassen, B. (2008). *Scientific representation: paradoxes of perspective.* New York: Oxford University Press.

Vandenbroucke, J. P. (2008). Observational research, randomised trials, and two views of medical science. *PLoS Med, 5*(3), e67.

Vrecer, M., Turk, S., Drinovec, J., & Mrhar, A. (2003). Use of statins in primary and secondary prevention of coronary heart disease and ischemic stroke: meta-analysis of randomized trials. *Int J Clin Pharmacol Ther, 41*(12), 567–77.

Wakefield, J. C. (1992). Disorder as harmful dysfunction: a conceptual critique of DSM-III-R's definition of mental disorder. *Psychol Rev, 99*(2), 232–47.

Walsh, D. (2015). *Organisms, agency, and evolution.* Cambridge: Cambridge University Press.

Welch, G. (2016). *Less medicine, more health.* Boston: Beacon Press.

West, S., King, V., Carey, T. S., Lohr, K. N., McKoy, N., Sutton, S. F., & Lux, L. (2002). Systems to rate the strength of scientific evidence: summary. AHRQ Evidence Report Summaries.

Whitaker, R. (2010). *Anatomy of an epidemic: magic bullets, psychiatric drugs, and the astonishing rise of mental illness in America.* New York: Crown.

Whitlock, M., & Schluter, D. (2009). *The analysis of biological data.* Greenwood Village: Roberts and Company Publishers.

Whittington, C. J., Kendall, T., Fonagy, P., Cottrell, D., Cotgrove, A., & Boddington, E. (2004). Selective serotonin reuptake inhibitors in childhood depression: systematic review of published versus unpublished data. *Lancet, 363*(9418), 1341–5.

Wieseler, B., Wolfram, N., McGauran, N., Kerekes, M. F., Vervölgyi, V., Kohlepp, P., ... Grouven, U. (2013). Completeness of reporting of patient-relevant clinical trial outcomes: comparison of unpublished clinical study reports with publicly available data. *PLoS Med, 10*(10), e1001526.

Wilholt, T. (2009). Bias and values in scientific research. *Studies in History and Philosophy of Science Part A, 40*(1), 92–101.

Wilholt, T. (2012). Epistemic trust in science. *The British Journal for the Philosophy of Science, 64*(2), 233–53.

Wilkinson, R. (2006). *The impact of inequality: how to make sick societies healthier.* New York: The New Press.

Wilson, M. C., Hayward, R. S., Tunis, S. R., Bass, E. B., & Guyatt, G. (1995). Users' guides to the medical literature. VIII. How to use clinical practice guidelines. B. What are the recommendations and will they help you in caring for your patients? The Evidence-Based Medicine Working Group. *JAMA, 274*(20), 1630–2.

Woodward, J. (2010). Causation in biology: stability, specificity, and the choice of levels of explanation. *Biology & Philosophy, 25*(3), 287–318.

Wootton, D. (2006). *Bad medicine: doctors doing harm since Hippocrates.* New York: Oxford University Press.

Worrall, J. (2002). What evidence in evidence-based medicine? *Philosophy of Science, 69,* S316–30.

Worrall, J. (2007). Why there's no cause to randomize. *The British Journal for the Philosophy of Science, 58*(3), 451–88.

Worrall, J. (2010). Do we need some large, simple randomized trials in medicine? In M. Suarez, M. Dorato, & M. Redei (eds.), *EPSA Philosophical Issues in the Sciences.* Dordrecht: Springer Netherlands.

Wylie, A. (1992). The interplay of evidential constraints and political interests: recent archaeo-logical research on gender. *American Antiquity, 57*(1), 15–35.

Yank, V., Rennie, D., & Bero, L. A. (2007). Financial ties and concordance between results and conclusions in meta-analyses: retrospective cohort study. *BMJ, 335*(7631), 1202–5.

Zhang, L., Steinmaus, C., Eastmond, D. A., Xin, X. K., & Smith, M. T. (2009). Formaldehyde exposure and leukemia: a new meta-analysis and potential mechanisms. *Mutat Res, 681*(2–3), 150–68.

General Index

Index of Drugs